Monographs on Theoretical and Applied Genetics 21

Monographs on Theoretical and Applied Genetics

Volume 10 **Male Sterility in Higher Plants**
By M.L.H. Kaul (1988)

Volume 11 **Tree Breeding: Principles and Strategies**
By G. Namkoong, H.C. Kang, and J.S. Brouard (1988)

Volume 12 **The Wheat Rusts – Breeding for Resistance**
By D.R. Knott (1989)

Volume 13 **Population Dynamics and the Tribolium Model Genetics and Demography**
By R.F. Costantino and R.A. Desharnais (1991)

Volume 14 **Genetic Improvement of Tomato**
Editor: G. Kalloo (1991)

Volume 15 **Evolution and Function of Heterostyly**
Editor: S.C.H. Barrett (1992)

Volume 16 **Distant Hybridization of Crop Plants**
Editors: G. Kalloo and J.B. Chowdhury (1992)

Volume 17 **Cytogenetics in Genetics and Plant Breeding**
By J. Sybenga (1992)

Volume 18 **Cytogenetics of the Festuca – Lolium Complex**
By P.P. Jauhar (1993)

Volume 19 **Breeding Oilseed Brassicas**
Editors: K.S. Labana, S.S. Banga, and S.K. Banga (1993)

Volume 20 **Oenothera**
By C. Harte (1994)

Volume 21 **Soil Mineral Stresses**
Editors: A.R. Yeo and T.J. Flowers (1994)

A.R. Yeo T.J. Flowers (Eds.)

Soil Mineral Stresses

Approaches to Crop Improvement

With 6 Figures

Springer-Verlag
Berlin Heidelberg GmbH

Dr. ANTHONY R. YEO
Dr. TIMOTHY J. FLOWERS

School of Biological Sciences
The University of Sussex
Brighton BN1 9QG, UK

ISBN 978-3-642-84291-7 ISBN 978-3-642-84289-4 (eBook)
DOI 10.1007/ 978-3-642-84289-4

Library of Congress Cataloging-in-Publication Data. Soil mineral stresses: approaches to crop improvement/A.R. Yeo, T.J. Flowers (eds.). p. cm. – (Monographs on theoretical and applied genetics; 21). Includes bibliographical references and index.

1. Crops – Genetic engineering. 2. Plant breeding. 3. Crops – Effect of minerals on. 4. Plant-soil relationships. 5. Crop improvement. I. Yeo, A.R. II. Flowers, T.J. (Timothy J.) III. Series. SB123.57.S65 1994 631.5'23—dc20 93-46743

Originally published by Springer-Verlag Berlin Heidelberg New York in 1994
Softcover reprint of the hardcover 1st edition 1994

Typesetting: Macmillan India Ltd., Bangalore-25

SPIN: 10024030 31/3130/SPS 5 4 3 2 1 0 – Printed on acid-free paper

Preface

This book is concerned with the ways in which crops might be developed for soils that are, at the present, agriculturally unproductive because of excesses and/or deficiencies of certain minerals. We have concentrated on soils, rather than on environmental conditions that limit productivity, since there have been a number of recent texts dealing with topics such as drought and temperature. The aim is that those working to derive crops for growth on these "problem" soils should be aware of the many diverse avenues that are available. These come from the disciplines of plant breeding, genetics and physiology, and the interfaces that are developing between them.

The background, the requirements to feed the projected increase in human population, is set out in the introductory chapter. The next five chapters then deal with the approaches to crop improvement: the merits of a conventional breeding programme, the importance of physiological characters in making selections, the use of in vitro techniques, of cytogenetics, and the value of developing native plants into crops in their own right. Although salinity is often used as an example, reflecting the research interests of many of the authors, the methods and approaches described have much wider applicability. Two chapters are then concerned more specifically with breeding for tolerance to other metal toxicities and with deficiencies and toxicities of micronutrients. Finally, in the concluding chapter, we summarise and find common ground between the different approaches and points of view.

Brighton, April 1994

ANTHONY R. YEO
TIMOTHY J. FLOWERS

Contents

Chapter 1 Introduction: World Population and Agricultural Productivity
T. J. Flowers

1.1	How Many People Are There?	1
1.2	Agricultural Requirements.	2
1.2.1	Food Requirements.	2
1.2.2	Food Supply.	4
1.3	Population-Carrying Capacity	5
1.4	How Much Land Is There?	6
1.5	Increasing Food Production	7
	References	9

Chapter 2 Conventional Plant Breeding for Tolerance to Problem Soils
C. N. Chaubey and D. Senadhira

2.1	Screening Techniques	11
2.2	Variability in Tolerance for Soil Stresses	14
2.2.1	Rice	15
2.3	Genetics of Tolerance for Soil Stresses	16
2.3.1	Gene Location and Linkages	19
2.3.2	Correlated Changes	20
2.4	Crop-Improvement	21
2.4.1	Introduction	21
2.4.2	Pure Line and Mass Selection	21
2.4.3	Hybridisation and Selection.	23
2.4.3.1	Pedigree Method	23
2.4.3.2	Bulk Method	23
2.4.3.3	Backcross Breeding.	26
2.4.3.4	Recurrent Selection.	26
2.4.3.5	Rapid Generation Advance Procedures	27
2.4.4	Mutation Breeding	28
2.4.5	Polyploid Breeding	28
2.4.6	Heterosis Breeding	29
2.5	Summary.	29
	References	29

Chapter 3 Physiological Criteria in Screening and Breeding
A. R. Yeo

3.1 Introduction . . . 37
3.2 Reasons for the Use of Physiological Selection . . . 38
3.2.1 The Complexity of Tolerance . . . 38
3.2.2 Importing Tolerance from Wild Relatives . . . 39
3.2.3 Measuring Stress Tolerance . . . 40
3.2.4 Interaction Between Environmental Stresses . . . 42
3.2.5 Limitations to the Use of Yield as a Selection Criterion . . . 42
3.3 Salinity . . . 43
3.3.1 Basic Problems . . . 43
3.3.2 Salt Exclusion . . . 44
3.3.3 Conditions Requiring Osmotic Adjustment . . . 45
3.3.4 Characteristics Needed in Salt-Tolerant Plants . . . 46
3.3.4.1 Control of Salt Uptake . . . 46
3.3.4.2 Limiting the Damage That Excessive Ion Uptake Causes . . . 46
3.3.4.3 Osmotic Adjustment . . . 47
3.3.5 Agricultural Versus Ecological Advantage . . . 47
3.3.6 Examples Where Knowledge of Physiological Mechanisms Would Aid Selection for Salt Tolerance . . . 48
3.3.6.1 Where Tolerance Is Accidental . . . 48
3.3.6.2 Where Tolerance Is Not a Single Character . . . 48
3.3.6.3 Where Tolerance Is Incidental . . . 49
3.3.6.4 Where a Mechanism Is As Important As Tolerance Itself . . . 49
3.3.6.5 Where Screening/Selection Pressure Is Severe . . . 49
3.4 Drought . . . 49
3.4.1 Efficient Use of Water . . . 50
3.4.2 Exploitation of Soil Moisture . . . 51
3.4.3 Leaf Water Relations . . . 52
3.4.4 Selection for Drought Tolerance . . . 53
3.5 Physiological Selection Procedures . . . 54
3.5.1 Advantages of Using Physiological Criteria . . . 55
3.5.2 Constraints . . . 55
3.5.3 Prospects . . . 55
References . . . 57

Chapter 4 Cytogenetic Manipulations in the Triticeae
B. P. Forster. With 5 Figures

4.1 Introduction . . . 61
4.2 Background to Wheat and Salinity . . . 61
4.3 Cytogenetics of Wheat . . . 63

4.4 Transfer into Wheat of Alien Genes for Tolerance to Salt. . 66
4.5 The Development of Hybrids as New Crop Species. 71
4.6 The Interface Between Cytogenetics and Physiology 71
4.7 Genetic Control of Salt Tolerance in Barley 73
4.8 Genes for Abiotic Stress Tolerance in the Triticeae 75
4.9 Examples of Alien Introduction 75
4.10 Genetic Markers in Plant Breeding 77
4.11 Summary. 78
References 78

Chapter 5 Tissue Culture in the Improvement of Salt Tolerance in Plants
P. M. Hasegawa, R. A. Bressan, D. E. Nelson, Y. Samaras, and D. Rhodes.
With 1 Figure

5.1 Introduction. 83
5.2 Application of Tissue Culture to Obtain Salt-Tolerant Plants 84
5.2.1 Somatic Cell Selection 84
5.2.2 Parasexual Hybridisation 89
5.2.3 Gene Transformation 91
5.3 Tissue Culture in the Identification and Characterisation of Cellular Determinants of Salt Tolerance. 93
5.3.1 Osmotic Adjustment Is a Fundamental Cellular Determinant of Salt Tolerance 94
5.3.2 Osmotic Adjustment Mediated by Ion Accumulation. . . . 94
5.3.3 Organic Osmotic Solute Accumulation in the Cytosol . . . 99
5.3.3.1 Glycinebetaine Accumulation. 101
5.3.3.2 Proline Accumulation 104
5.3.3.3 Accumulation of Other Organic Solutes 109
5.4 Conclusion. 110
References 111

Chapter 6 The Agricultural Use of Native Plants on Problem Soils
J. W. O'Leary

6.1 Evolution of Domestic Species 127
6.2 Limits of Improving Existing Crops 127
6.2.1 Genetic Variability Within Crop Species 127
6.2.2 Salt Tolerance. 128
6.2.3 Mineral Deficiency 128
6.2.4 Heavy Metal Toxicity 129
6.2.4.1 Extent of the Resource Base 130
6.2.4.2 Heavy Metal-Binding Proteins 130
6.3 Availability of Alternatives 131

6.4 Methods of Domestication 131
6.4.1 Extent of the Halophyte Resource Base. 131
6.4.2 Use of Halophytes to Improve Rangeland Productivity . . 132
6.4.3 Use of Halophytes as Irrigated Crop Plants 133
6.4.3.1 Use of Forage Crops: Productivity. 133
6.4.3.2 Potential for Use as Direct Food Crops 135
6.5 Time Scale for Agricultural Development. 137
6.6 Conclusions 139
References 140

Chapter 7 Metal Toxicity
T. McNeilly

7.1 Introduction: Sources of Toxicity 145
7.2 The Measurement of Tolerance. 147
7.2.1 Solution Culture Methods. 148
7.2.2 Soil-Based Methods of Testing Tolerance. 150
7.3 Variability in Wild Species 152
7.3.1 Tolerance to Heavy Metals 152
7.3.2 Tolerance of Aluminium and Manganese 153
7.4 Variability in Cultivated Species 154
7.4.1 Tolerance to Zinc. 155
7.4.2 Tolerance to Aluminium and Manganese. 155
7.4.2.1 Wheat 155
7.4.2.2 Sorghum 157
7.4.2.3 Rice 157
7.4.2.4 Maize. 157
7.4.2.5 Other Species 158
7.5 The Genetic Basis of Metal Tolerance. 159
7.5.1 Wild Species. 159
7.5.2 Cultivated Species 160
7.6 The Physiological Basis of Tolerance to Metals 164
7.7 Synthesis 165
References 168

Chapter 8 Micronutrient Toxicities and Deficiencies in Rice
H. U. Neue and R. S. Lantin

8.1 Introduction 175
8.2 Iron 176
8.2.1 Iron Deficiency 178
8.2.2 Iron Toxicity 180
8.3 Manganese. 182

8.3.1 Manganese Deficiency 183
8.3.2 Manganese Toxicity 184
8.4 Zinc. 185
8.4.1 Zinc Deficiency 186
8.5 Copper 189
8.5.1 Copper Deficiency 190
8.6 Boron. 191
8.6.1 Boron Deficiency 191
8.6.2 Boron Toxicity 192
8.7 Molybdenum 192
References 193

Chapter 9 Summary: Breeding Plants for Problem Soils – Current Knowledge and Prospects
A.R. Yeo and T.J. Flowers

9.1 Why Grow Crops on Problem Soils? 201
9.2 Approaches to the Utilisation of Problem Soils 202
9.3 Selection of Parents and Within Breeding Populations . . . 203
9.4 Tolerance and Potential Yield 206
9.5 Genetics of Tolerance to Problem Soils. 207
9.6 Transfer from Other Species 208
9.7 Domestication of New Crops from the Native Flora 210
9.8 Outlook 211

Subject Index 213

List of Contributors

Bressan, R.A.
Department of Horticulture, Purdue University, West Lafayette, Indiana 47907, USA

Chaubey, C.N.
International Rice Research Institute, P.O. Box 933, 1099 Manila, The Philippines

Flowers, T.J.
School of Biological Sciences, The University of Sussex, Brighton BN1 9QG, UK

Forster, B.P.
Scottish Crop Research Institute, Invergowrie, Dundee DD2 5DA, Scotland

Hasegawa, P.M.
Department of Horticulture, Purdue University, West Lafayette, Indiana 47907, USA

Lantin, R.S.
International Rice Research Institute, P.O. Box 933, 1099 Manila, The Philippines

McNeilly, T.
Department of Environmental and Evolutionary Biology, School of Life Sciences, The University of Liverpool, P.O. Box 147, Liverpool L69 3BX, UK

Nelson, D.E.
Department of Horticulture, Purdue University, West Lafayette, Indiana 47907, USA

Neue, H.U.
International Rice Research Institute, P.O. Box 933, 1099 Manila, The Philippines

O'Leary, J.W.
Bioresources Research Facility, Office of Arid Zone Studies,
University of Arizona, 250 East Valencia Road, Tucson,
Arizona 85706, USA

Rhodes, D.
Department of Horticulture, Purdue University, West Lafayette, Indiana 47907,
USA

Samaras, Y.
Department of Horticulture, Purdue University, West Lafayette, Indiana 47907,
USA

Senadhira, D.
International Rice Research Institute, P.O. Box 933, 1099 Manila,
The Philippines

Yeo, A.R.
School of Biological Sciences, The University of Sussex, Brighton BN1 9QG,
UK

Chapter 1

Introduction: World Population and Agricultural Productivity

T.J. FLOWERS

1.1 How Many People Are There?

For most of the 16000 years of human history, there have been less than 300 million people on the earth, but in the last 200 years a sustained and rapid increase in population has taken place (Salk and Salk 1981). The number passed 1 billion (1×10^9) in 1850, 2.5 billion a century later, and the remainder of the 20th century will see the population rise from the current level of about 5.3 billion to over 6 billion (Table 1.1). In the longer term, it is predicted that the human population will reach a distinct plateau (Salk and Salk 1981); these predictions are based on the change in populations in the more developed countries and analysis of animal populations. This plateau will occur at a value of between about 10 and 14 billion people (see Sadik 1990).

The world's population and its rate of growth are distributed very unevenly. The global rate of population growth is at present 1.8% per annum but it is very different in developing and developed countries. Annual growth rates in developing countries reached a maximum of about 2.4% in the decade 1960–70 and this is predicted to drop to 1.6% by the end of this decade (Department of International Economic and Social Affairs 1980), while in the more developed countries the peak growth rate of 1.3% was reached in 1950 and should drop to 0.5% by the year 2000 (see Salk and Salk 1981). This pattern is largely a result of an imbalance between changes taking place in birth and death rates; the decline in death rate precedes and is greater than the decline in the birth rate; but as economies develop, birth rates do tend to decline and populations go through a so-called demographic transition (see, for example, Findlay and Findlay 1987). The contrast between the developed and developing world is highlighted by the calculated doubling times for the populations of various countries: they range from 18 years (Kenya) to 58 years (China) in the less developed, and from 86 years (in what was the USSR) to 1155 years (in Sweden) in the more developed, countries (Salk and Salk 1981). In some countries of Europe, (e.g. W. Germany), growth rates have fallen below replacement rates. As far as the world as a whole is concerned, however, such countries are, unfortunately, the exception.

By the turn of the millennuim, about 80% of the people will be in what are now the less developed countries, and half of the population are likely to be urban dwellers (see Tabah 1982). Ninety-five percent of the increase in

Monographs on Theoretical and Applied Genetics, Vol. 21
Ed. by A. R. Yeo and T. J. Flowers

Table 1.1. The estimated human population of the world between 1950 and 2025

Year	World population (billion)		Proportion in less developed countries (%)[b]
1950	2.5[a]		
1980	4.4		74.5
1990	5.2	5.3[c]	77.0
2000	6.1	6.25	79.2
2010	7.0		81.1
2025	8.2	8.5	83.2

[a] From the Department of International Economic and Social Affairs (1980).
[b] Tabah (1982).
[c] Sadik (1990).

population over the next 35 years will take place in Africa, Asia and Latin America (Sadik 1990). A vital challenge to those concerned with agriculture will be to maintain the food production at a level that does not reduce food supply on a per capita basis while the population increases according to the projections in Table 1.1.

1.2 Agricultural Requirements

1.2.1 Food Requirements

This century's rapid population growth has been accompanied by a dramatic increase in the production of food, as exemplified by the increase in the amount of grain grown in the world (Table 1.2). As a global average, food production has, to date, kept pace with population growth and there has been a small increase in the food production per head of the population. However, as with population growth, there are wide differences between the developing and developed world in both food consumption and food production. The peoples in the countries of the developing world obtain on average only about 95% of their estimated calorie requirements compared with 130% of the requirement in the developed world (Department of International Economic and Social Affairs 1980). Averages, of course, disguise much greater local imbalances.

The production of major food crops per head of the population is about three times higher in the developed than in the developing nations (Table 1.3). The developing nations, with about 75% of the world's population, generate less than half of the world's major crop production. Furthermore, although the growth in the production of food has, since the 1960s, exceeded the growth in population, the differential has been decreasing – that is the growth of food

Table 1.2. World population, population change and production of grain. (Brown 1984)

Year	Population (millions)	Grain	
		total ($\times 10^6$ tonnes)	Per capita (kg)
1950	2510	631	251
1960	3030	863	285
1970	3680	1137	309
1980	4420	1432	324

Table 1.3. Food production in developing and developed nations; 1976–1980 averages. (Mellor and Paulino 1986)

Area	Population ($\times 10^6$)	Major food crop production	
		Total ($\times 10^6$ tonnes)	Per capita (kg)
Developed	1147	851	740
Developing	3111	758	240

production per head of the population has been in decline (Department of International Economic and Social Affairs 1980), and cereal production per person during the 1980s increased in only 43 developing countries, declining in 51 others (Sadik 1990).

Based upon the trends over the period 1961–1977, food production in the developing countries would rise from 758×10^6 tonnes (Table 1.3) to 940×10^6 tonnes in the year 2000. At this time there would be a shortfall of production compared with demand of 80×10^6 tonnes (Mellor and Paulino 1986; Table 1.4), about four times the deficit in 1977. However, it is no foregone conclusion that the upward trends in the area of land cultivated, and unit area productivity (brought about by plant breeding or husbandry), will continue, and the deficit may be greater. Of the developing areas, only Asia is projected to meet its demands. There will be major deficits in North Africa and the Middle East and Sub-Saharan Africa: Latin America will have a small deficit (Table 1.4). In Latin America the major problem may be one of food distribution between rich and poor, while in Sub-Saharan Africa the need will be to re-vitalise agriculture (Toenniessen 1984).

The process of economic development and increase in per capita income is also a key issue in the world's food problem. Pressures on food production are not simply driven by increases in population but also by the wealth of the population. Increases in wealth do change dietary habits and lead, for instance, to greater demands for meat which can mean a consequent lengthening of the

Table 1.4. Projected food production and consumption for developing countries in the year 2000. Figures are in millions of tonnes. The analysis assumes that the trend in income growth continues, as well as the demand that such growth brings. (Mellor and Paulino 1986)

Country	Production	Consumption	Surplus/deficit
Total	940	1020	− 80
Asia	410	463	+ 18
North Africa/Middle East	120	177	− 57
Sub-Saharan Africa	113	148	− 36
Latin America	227	232	− 5

food chain and decrease in the overall efficiency of production. For example, the necessary consumption of grain per head is about 180 kg per annum if eaten directly, but rises to 750 kg per annum if used to produce a meat-based diet (Freeman 1984). However, animals can be raised on land unsuited to arable agriculture and this would give a high protein product from land that will not support cereals.

1.2.2 Food Supply

Deficits in production in a given country or region mean that either people suffer deficiencies in their nutrition and/or the country must import food. The latter may be obtained either through the use of the country's wealth or through foreign aid. Between 1973 and 1977, developing countries imported more food than between 1961 and 1965 – 23×10^6 as opposed to 5.5×10^6 tonnes – a consequence of decline in production and an increase in demand. Countries with a very slow economic growth rate (an increase in GNP per capita of less than 1%) have a particularly high and increasing food import, based on foreign aid (such changes are documented in Mellor and Paulino 1986). Overall estimates of the demand for food in the world suggest a growth of 3–4% per annum (Toenniessen 1984). There are two ways in which such an increase in food production might be effected: an increase in the area under production or an increase in the productivity per unit area. The former depends on the availability of land and the latter on the interaction between the biology of yield and resource inputs.

Up to the middle of this century, most of the rise in food production had come from an increase in the area under cultivation. Between 1950 and the present, however, the rise has increasingly come from growth in yields per unit area farmed (Brown 1984). Between 1961 and 1977, more than 60% of the average increase of 2.6% per annum in the production of major food crops in the developing nations (Table 1.5) was attributable to a rise in output per hectare.

A worrying feature of recent figures, however, is that the growth rate of production was less in the 1970s as compared with the 1960s, due to a decline in

Table 1.5. Average annual growth rates of production, area harvested and output per hectare (%) in developing countries. (Mellor and Paulino 1986)

Period	Production	Area harvested	Output per ha	Relative contribution	
		(%)		Area	Output
1961–77	2.6	1.0	1.6	38	62
1961–69	2.6	1.4	1.2	53	47
1970–77	2.4	0.9	1.5	37	63

the growth rate of the area harvested. This was largely due to a substantial decline in Sub-Saharan Africa and Latin America. Loss of once-productive land contributes to this and the figures suggest a lack of sustainability of some forms of agricultural practice – of land degradation through erosion, loss of fertility and salinisation, for example. The challenge to meeting food requirements is not simply a question of increasing production at whatever cost. Emphasis must inevitably centre on long-term sustainability: how much land can be kept productively in agriculture and what yield can be reliably maintained?

1.3 Population-Carrying Capacity

In a detailed survey of the potential population-carrying capacity of the developing world, the Food and Agriculture Organization of the United Nations in collaboration with its Fund for Population Activities and the International Institute for Applied Systems Analysis evaluated, during the late 1970s, the potential productivity of land in several regions: Africa, Southwest Asia, South America, Central America and Southeast Asia. The model assumed that each country is an entity with free movement of crops within its boundaries, and did not take account of any increase in demand through increased wealth. The potential calorie production from 18 crops suitable for the particular climatic conditions was calculated under three levels of inputs and for the populations existing in 1975 and projected for the year 2000 (Higgins et al. 1982). The authors concluded that the lands of the developing world (excluding East Asia) were able to support twice the population present in 1975 and 1.5 times the projected population of the year 2000. The improvements in the potential population-carrying capacity between 1975 and 2000 were due to projected increases in production under irrigation, taking account of loss of land through increased population pressure. There was large regional variation, however, with the poorest prognosis being for Southwest Asia (the countries often now described as the Middle East, from Turkey to Afghanistan to Saudi Arabia), where only with the highest level of inputs was the productivity adequate to meet projected growth of population.

High inputs assumed complete mechanisation, the full use of the best crop varieties (those with maximum calorie-protein content) and application of necessary chemicals and soil conservation measures. If this could be supported (presumably through aid), it presents a disturbing scenario of industrialisation, urbanisation and the widespread breakdown of existing agriculture-based societies. If only low levels of inputs are available – hand labour, no fertiliser or pesticides and no soil conservation measures with the present crops grown on potentially cultivatable rainfed lands – 65 of the 117 countries included in the survey would be unable to meet the demands of their populations by the year 2000.

1.4 How Much Land Is There?

The land surface of the world (about 29% of the globe) is nearly 150 million km^2 (Rodin et al. 1975). One tenth of this (1.49 billion ha in 1976; Department of International Economic and Social Affairs 1980) is currently arable (cultivated land under permanent crop production). Between 1870 and 1970, the amount of land used in the world as arable approximately doubled. From 1970 to 1980, the increase in arable land in the developed world was some 0.5%, whereas it was about 5% in the developing world (Mather 1986).

There are widely conflicting views of the area that could be arable and upper estimates of the amount suitable for cultivation range from 3.2 to 5.0 billion ha (Table 1.6). However, the consequences of wholesale removal of native vegetation are now too well documented to advocate the expansion of arable land as a simple answer to the problem of increased food production. Increasing population and increasing wealth will also place demands on land use for purposes other than agriculture. It seems likely that the rate of increase of arable land will decline over the coming decades in comparison with that of the last 30 years. It must also be borne in mind that most of the good arable land is

Table 1.6. The extent of land and agricultural land

	(billions of ha)
World land surface	14.93[a]
Agricultural land	1.488[b]
Developed	0.673[b]
Developing	0.815[b]
Suitable for cultivation	3.2[c]
	1–5[d]

[a] From Rodin et al. (1975). [b] Department of International Economic and Social Affairs (1980). [c] Times Atlas (1980). [d] FAO (1978).

already in use. On the whole, new land that is brought into cultivation is likely to have some associated problem, such as poor fertility, poor drainage, liability to drought, the presence of salinity or alkalinity. This is usually the reason that it has not previously been used, and crop yields on such land are likely to be lower than the average of present arable. It is, however, exactly for these soils that plant breeders must develop crops, the cultivation of which will be sustainable.

As mentioned earlier, an important aspect of land use is the extent to which land is lost to agriculture through such processes as erosion, waterlogging, salinisation, contamination by toxic metals and desertification. The areas affected are not easily quantifiable (Mather 1986), but are clearly important in agriculture – particularly because they represent a loss of previously productive land. An example of the magnitude of the problem can be seen in figures quoted for the extent of salinisation of irrigated land – between one third and one half is saline (Wittwer 1979; Kovda 1980). These areas constitute "problem soils" and crops need to be developed to help restore their agricultural potential.

Although the discussion so far has centred on cultivation, not all land that might be used for agriculture is suitable for arable. There are large tracts of the world's surface (40% of agricultural land, according to Wittwer 1986) which are not suitable for cultivating crops, but which can be efficiently used by animals. Some of these areas may be subject to soil problems and be those for which the development of the native flora is important.

It does not, in summary, appear that food supply could be increased in line with the projected increases in human population through increases in the area of land cultivated without increases in productivity per unit area.

1.5 Increasing Food Production

There is clearly a large potential for improvement in the biological sense. The extent to which yield might, theoretically, be increased is suggested by data on maize, wheat and barley from the US. Here the best yields achieved by farmers are commonly two to three times the average yields, but still considerably lower than the record yields (Table 1.7). If there are such dramatic differences between actual and potential yields (three- to sixfold: Table 1.7; calculations suggest that the potential for global food production is some 40 times the present level: see Wittwer 1986) why the concern about shortfalls in food production? A 10% increase in yields would compensate for the demands highlighted in Table 1.4.

The best and record yields (Table 1.7) are, however, obtained on ideal soils and conditions with very high levels of inputs (fertilisers and husbandry such as mechanisation and pesticides). These yields could not be translated to all soils and climates and the inputs could not be achieved, let alone sustained, by farmers in developing countries.

In the short term, the quantity of food required to feed a given population can be, and has sometimes been, met without regard to whether the production

Table 1.7. Average, best and record yields for crops in the United States of America in 1979. (Wittwer 1986)

Food crop	Crop yield		
	Average	Best	Record
	(tonnes/ha)		
Maize	6.9	14.7	22.2
Wheat	2.3	6.9	14.5
Sorghum	4.0	17.8	21.5
Potato	31	68	95
Sugarcane	84	140	250

can be sustained – sufficiency has, in some instances, been achieved at the expense of sustainability (the concept of sustainability is discussed in a series of essays in Douglass 1984).

More intensive production on land already used for agriculture will, however, be a necessary and major part of the way to increase further food supply. Land that has been damaged or lost through agriculture cannot be "written off". Breeding plants for productivity on poor soils will not have the same aims in terms of yield per unit area as that on highly fertile areas. For poor and problem soils, it will be important to understand the constraints on production and to evaluate whether or not they can be overcome – and if so by what means, biological or technological? This will apply both to the development of new lands and to attempts to reinstate agricultural productivity on degraded land.

Decisions in plant breeding programmes involve decisions at various levels – whether a biological increase in yield can be accomplished, whether such an increase has environmental costs which society will see as valid, whether society is already sufficient in the commodity, whether the costs of the breeding programme are justified in the returns to be achieved (and how broadly those returns are calculated; e.g. are environmental costs taken into consideration or merely the profits to be made by a company marketing seed, fertiliser or insecticides?). There are, thus, scientific, political, social and economic aspects to any decisions. It is with the biological aspects of the problem that this book is concerned (but the other aspects should not be ignored).

It is also clear that, to date, the majority of effort in plant breeding has concerned but a few plant species and mainly those of high economic value. Commercial pressures may dictate that this trend continues into the future. For the rural poor, however, the market value of a crop may be relatively unimportant – the crop, the product of their labour, is for their own consumption, not for sale. There are a number of crops that are especially important in this respect (e.g. legumes) and for which no large research organisation has taken particular responsibility. It is also clear that humans use a very small proportion of the plant species available as crops – perhaps 230 (Simmonds 1976) of the 250 000 known species (Heywood 1985): just 15 species dominate world trade. Diversity may be particularly important in any system of sustainable agriculture that operates in the future.

References

Brown LR (1984) World population growth, soil erosion, and food security. In: Douglass GK (ed) Agricultural sustainability in a changing world order. Westview, Boulder, pp 31–48

Department of International Economic and Social Affairs (1980) World population trends and policies 1979. Monitoring Report, vol 1. Population trends. Population studies 70. United Nations, New York

Douglass GK (ed) (1984) Agricultural sustainability in a changing world order. Westview, Boulder

FAO (1978) How much good land is left? CERES 13: 12–16

Findlay A, Findlay A (1987) Population and development in the third world. Methuen, London

Freeman OL (1984) A global strategy for agriculture. In: Douglass GK (ed) Agricultural sustainability in a changing world order. Westview, Boulder, pp 135–143

Heywood VH (ed) (1985) Flowering plants of the world. Croom Helm, London

Higgins GM, Kassam AH, Naiken L, Fischer G, Shah MM (1982) Potential population supporting capacities of lands in the developing world. Technical Report of Project INT/75/P13 'Land Resources for Populations of the Future'. Food and Agriculture Organization of the United Nations, Rome

Kovda VA (1980) Land aridization and drought control. Westview, Boulder

Mather AS (1986) Land use. Longman, London

Mellor JW, Paulino L (1986) Food production needs in a consumption perspective. In: Swaminathan MS, Sinha SK (eds) Global aspects of food production. Tycooly International, Oxford, pp 1–24

Rodin LE, Bazilevich NI, Rozov NN (1975) Productivity of the world's main ecosystems. In: Productivity of world ecosystems. National Academy of Sciences, Washington, pp 13–26

Sadik N (1990) The state of world population 1990. UNFPA. United Nations Population Fund, New York

Salk J, Salk J (1981) World population and human values – a new reality. Harper and Row, New York

Simmonds NW (ed) (1976) Evolution of crop plants. Longman, London

Tabah L (1982) Population growth. In: Faaland J (ed) Population and the economy in the 21st century. Blackwell, Oxford, pp 175–205

Times Atlas (1980) Times atlas of the world comprehensive edition. Times Publishing, London

Toenniessen GH (1984) Review of the world food situation and the role of salt-tolerant plants. In: Staples RC, Toenniessen GH (eds) Salinity tolerance in plants – strategies for crop improvement. Wiley, New York, pp 399–413

Wittwer SH (1979) Future technological advances in agriculture and their impact on the regulatory environment. Bioscience 29: 603–610

Wittwer SH (1986) Research and technology needs for the twenty-first century. In: Swaminathan MS, Sinha SK (eds) Global aspects of food production. Tycooly International, Oxford, pp 85–116

Chapter 2

Conventional Plant Breeding for Tolerance to Problem Soils

C. N. CHAUBEY and D. SENADHIRA

Crop cultivars grown at present on problem soils have been selected or developed by conventional methods of plant breeding. The choice of a suitable breeding programme for the development of a tolerant cultivar to a defined soil stress depends upon a number of factors: screening techniques, sources and mechanisms of tolerance, genetic variability, modes of gene action and heritability, and their relationship to agronomic traits.

2.1 Screening Techniques

A large number of techniques have been used by researchers to isolate genotypic differences for tolerance to various soil-related stresses. They have been used for identifying tolerant genotypes, for genetic and physiological investigations, and for breeding. These techniques can be classified broadly under four headings: laboratory, greenhouse, artificial field and natural field.

Laboratory techniques are usually used for screening at germination or the early vegetative stages of growth. Pot culture and solution culture are common greenhouse techniques. Determination of tolerance throughout the growth cycle is feasible for most crop species using greenhouse techniques, although capacity may be limiting in practice. Screening in the field, where stress is artificially induced, is widely practiced. For mass-scale screening, such as for early generation breeding materials, fields with naturally occurring stress are the most economical. There are difficulties, however, in field situations due to lack of uniformity of the stress and to the occurrence of other stresses. In general, it is difficult to separate tolerance to one particular stress in field screening. Different techniques are applicable to different problems, and the discussion which follows concentrates on rice as an example.

Laboratory, greenhouse and field techniques used at the International Rice Research Institute (IRRI) for screening rice varieties against soil stresses are summarised in Table 2.1. Since the capacity of these techniques is limited to relatively small numbers of entries they are used only to screen varieties and advanced breeding lines. A range of methods has been developed elsewhere to screen rice varieties at seed germination, during seedling growth and at the mature stages of the life cycle. Salinity tolerance, for example, has been measured

Monographs on Theoretical and Applied Genetics, Vol. 21
Ed. by A. R. Yeo and T. J. Flowers

Table 2.1. Procedures used at IRRI for screening rice varieties against soil stresses. (R. S. Lantin, Soils and Water Science Division, IRRI)

Stress	Location/medium	Fertilizers	Method of planting	Layout	Check varieties (T = tolerant, S = sensitive)	Scoring system[a]
Salinity	Greenhouse, Maahas clay treated with common salt to an Ec of 8 dS/m	25 mg/kg as urea	Transplant 2-week-old seedlings grown in culture solution	Trays, 35 × 27 × 11 cm 3 seedlings/ variety 3 varieties/tray	T: Pokkali S: IR5929-12-3, M1-48 T and S checks after every 20 entries	SES at 4 WAT
	Field, Maahas clay treated with common salt to an Ec of 8 dS/m	50 kg N/ha as urea	Transplant 1 month-old seedlings grown on wet seedbed	2.1-m row, 20 × 15 cm	T: IR4630-22-2-5-1-3 IR9884-54-3 S: IR5929-12-3 T and S checks after every 18 entries	SES at 4 WAT
	Field, Maahas clay treated with common salt to an Ec of 8 dS/m	50 kg N/ha as urea	Transplant 1 month-old seedlings grown on wet seedbed	Three 5-m rows, 20 × 20 cm	T: IR4630-22-2-5-1-3 IR9884-54-3 randomized with entries S: IR5929-12-3 after every 10 entries	SES at 4 and 8 WAT
Alkalinity	Greenhouse, Maahas clay treated with Na_2CO_3 to obtain a pH of 8.6 SAR: 35	25 mg/kg as urea 4% zinc oxide dip	Transplant 2-week-old seedlings grown on culture solution	Trays, 35 × 27 × 11 cm 3 seedlings/ variety 3 varieties/tray	T: IR46, IR36 S: IR45 T and S checks after every 18 entries	SES at 4 WAT
	Field, Maahas clay treated with Na_2CO_3 to obtain a pH of 8.6 SAR: 35	50 kg N/ha a urea 4% zinc oxide dip	Transplant 1 month-old seedlings raised on wet seedbed	2.1 m rows, 20 × 15 cm	T: IR46 S: IR5931-110-1 T and S checks after every 18 entries	SES at 4 WAT
	Field, Maahas clay treated with Na_2CO_3 to obtain a pH of 8.6 SAR: 35	50 kg/ N/ha as urea 4% zinc oxide dip	Transplant 1 month-old seedlings raised on wet seedbed	Three 5-m rows, 20 × 20 cm	T: IR46 randomized with entries S: IR5931-110-1 after every 10 entries	SES at 4 WAT

Zinc deficiency	Field, pH: 6.7, Organic carbon: 3.9%, Avail Zn: 0.19 ppm	No fertilizer	Transplant 3-week-old seedlings grown in trays in greenhouse	Three 5-m rows 20 × 20 cm	T: IR54 randomized with entries S: IR5931-110-1 after every 10 entries	SES at 4 WAT
Phosphorus deficiency	Greenhouse, culture solution	P levels, 0.5 ppm and 10 ppm	Direct seed	3-l pots	T: IR54 S: IR6115-1-1 after every 50 entries	SES at 4 WAT
	Field, Luisiana clay, pH: 4.9 Organic carbon: 4.3%, Avail P: 3 mg/kg	50 kg N/ha as urea 25 kg K/ha as muriate of potash and 25 kg P/ha as single super-phosphate	Transplant 3-week-old seedlings grown on same soil	5-m rows, 20 × 20 cm	T: IR54 randomized with entries S: IR6115-1-1 after every 10 entries	SES at 4 or 6 WAT
Iron toxicity	Greenhouse, acid soil (pH = 3.7) containing more than 300 ppm water-soluble iron	50 mg N/kg as urea 25 mg P/kg as super-phosphate 25 mg K/kg as muriate of potash	Transplant 2-week-old seedlings grown on culture solution	16-l pots	T: IR4683-54-2-2-3 S: IR45 T and S checks after every 20 entries	SES at 4 or 6 WAT
Boron toxicity	Greenhouse, Maahas clay treated with 15 mg B/kg as borax	50 mg N/kg as urea 25 mg P/kg as single super-phosphate	Transplant 2-week-old seedlings grown on culture solution	16-l pots	T: IR9129-0-2-2-3 S: IR5929-1-2-3 T and S checks after every 20 entries	Scores at 8 WAT and maturity

[a] SES = standard evaluation system for rice (IRTP 1988); WAT = weeks after transplanting.

during germination in NaCl solutions (Shafi et al. 1970), mixed salt solutions, salinized sand culture (Pearson et al. 1966), or soil culture (Barakat et al. 1971) as well as in containers filled with salt-moistened filter papers (Younis and Hatata 1971) or salinized agar (Carlson et al. 1983). Seedlings have been screened in salinized culture solution (IRRI 1968; Akbar and Yabuno 1974; Yeo et al. 1990). Plants have been grown to maturity on saline soil in pots (Datta 1972), microplots (Janardhan and Murty 1972) or in the field (Purohit and Tripathi 1972).

Field evaluation of varieties for alkalinity tolerance has been conducted on alkali field soils by recording alkali injuries and grain yield (Purohit and Tripathi 1972). Field screening has also been used for P-efficiency in rice in the Gampola area of Sri Lanka (Ikehashi and Ponnamperuma 1978), where plants were grown on lateritic fields (terraces) either without added P or with P added at 112 kg P_2O_5/ha. Forty days after transplanting, lines were evaluated on the basis of tiller number. Other morphological markers which have been used for field P-efficiency screening in rice include the number of filled grains/plant or hill, and shoot dry weight of mature plants (Majumder et al. 1989; Chaubey and Senadhira 1991 unpubl.).

Fe-deficiency (regardless of soil pH) and manganese (Mn) and aluminium (Al) toxicities in acid soils are important mineral stresses that affect plant yield on aerobic soils. Screening procedures for tolerance to individual stresses involve direct sowing of the seeds of different rice entries on three different well-drained soils in concrete tanks and determining their grain yield at harvest. Later on, selected entries are re-evaluated under field conditions (IRRI 1972). Field screening has also been found to be effective for tolerance to acid sulphate soils and histosols (Ikehashi and Ponnamperuma 1978).

In rice, screening for Al tolerance in solution culture on the basis of root length of plants has been found to be effective since relative root length values were correlated with grain yield in the field (Howeler and Cadavid 1976; IRRI 1978; see also Chap. 7.5.2). A second technique used with rice for screening for tolerance to Al is root regrowth, which is the extent of root recovery after Al stress (Martinez 1976). Coronel et al. (1990) compared three techniques – hematoxylin staining (Polle et al. 1978), absolute root length and root regrowth – and concluded that relative root length was better than the other two.

The techniques described above are designed for single stresses in isolation and do not allow simultaneous screening of plants against several stresses. This is often necessary for selecting early generation breeding materials, and for this the technique used by breeders is to screen materials in target environments.

2.2 Variability in Tolerance for Soil Stresses

There is now ample evidence for the existence of inter- and intraspecific variability for tolerance to soil stresses. A number of articles and reviews on genetic variability for plant adaptability to edaphic stresses have encouraged scientists to intensify breeding programmes for developing mineral-efficient or toxicity-tolerant crop varieties (Myers 1960; Vose 1963; Antonovics et al. 1971; Brown et al. 1972; Epstein 1972; Läuchli 1976; Foy et al. 1978; Jung 1978; Clark and Brown 1980).

Most studies have been concerned with salinity, alkalinity, Fe deficiency/toxicity, Al toxicity, Mn toxicity, P deficiency and Zn deficiency. Variability has been widely observed within and between crop species. Tolerance for deficiencies and toxicities of several minerals have been found in soybean, cotton and sorghum germplasm (Brown and Jones 1977a, b, c). Barley, wheat and oats were tolerant, less sensitive and most sensitive, respectively, to Mn deficiency (Nyborg 1970). Oats had greater sensitivity to Fe-deficiency than wheat, rye, triticale and barley (McDaniel and Dunphy 1978). Many forage grasses and legumes such as *Hyparrhenia rufa*, *Melinis minutiflora* and *Brachiara decumbens* are tolerant of a high degree of soil acidity (Spain et al. 1975; Spain 1976). In *Trifolium* species, cultivars of berseem, red and rose clover exhibit tolerance to Fe-deficient soils but those of crimson, subterranean and arrowleaf clover show variable susceptibility (Gildersleeve and Ocumpaugh 1988). The threshold for an effect of salinity upon yield varies from 1 dS/m for beans to about 8 dS/m for barley, cotton and some *Agropyron* species (Rana 1986). A 50% yield reduction is caused by 18 dS/m in barley, cotton and certain other crops, but by only 3.6 dS/m in beans (Francois and Bernstein 1964; Maas and Hoffman 1976; West and Francois 1982; Maas 1987).

Three hundred cultivars of spring and winter wheat have been tested in a greenhouse for tolerance to highly acid soil conditions, and a wide range of variation observed (Mesdag and Slootmaker 1969). Wheat cultivars also differ in sensitivity to copper (Cu) deficiency (Hill et al. 1978). Maize genotypes, grown on a Zn-deficient soil, vary for Zn efficiency (Shukla and Raj 1976). Soybean cultivars vary in their tolerance to Zn toxicity (White et al. 1979) and Mn toxicity (Carter et al. 1975). Soybean cultivars of different maturity groups have been found to differ for tolerance to Al toxicity on an acid soil (Armiger et al. 1968). Abel and MacKenzie (1964) have reported wide variation for salt tolerance in soybean. Differential response to Mn toxicity has been found among potato cultivars (Ouelette and Genereux 1965). Foy et al. (1979) have observed several cotton genotypes to be tolerant to Al toxicity in an acid soil. In the case of grasses, genetic variability for Al tolerance has been found in Kentucky blue grass and fine-leaf fescue cultivars (Murray and Foy 1978) and several genotypes of *Macroptilium* forage legume also differ for Mn tolerance (Hutton et al. 1978). As reported by Dessureaux and Ouellette (1958), strains of alfalfa vary in their tolerance to Mn and Al toxicity.

2.2.1 Rice

There is considerable variability in tolerance for soil stresses among rice cultivars. Using laboratory, greenhouse, and field techniques of screening, varieties have been identified that possess tolerance to salinity (IRRI 1968;

Akbar and Yabuno 1974; Ikehashi and Ponnamperuma 1978; Yeo and Flowers 1984; Aslam 1989), alkalinity (Purohit and Tripathi 1972; Ikehashi and Ponnamperuma 1978), Al and Mn toxicity (IRRI 1970, 1978; Howeler and Cadavid 1976), Fe toxicity (Jayawardena et al. 1977; Virmani 1977; Ikehashi and Ponnamperuma 1978), P deficiency (IRRI 1971, 1976; Katyal et al. 1975; Ikehashi and Ponnamperuma 1978) and Zn deficiency (IRRI 1972, 1973, 1974, 1975, 1977, 1989). Up to December 1990, IRRI had screened about 185000 rice varieties and breeding lines for tolerance to salinity, alkalinity, peat soil conditions, Fe toxicity, P deficiency, Zn deficiency, Fe deficiency, Al and Mn toxicity and boron (B) toxicity. On the average, about 12% of the entries were classified as tolerant.

2.3 Genetics of Tolerance for Soil Stresses

The uptake, accumulation and utilisation of mineral elements in plants are genetically controlled, although there is a strong environmental interaction. Knowledge of the inheritance of tolerance for soil stresses helps in determining the most suitable procedure to be followed in breeding for tolerance to such stresses. Plants tolerant to mineral deficiency (also termed mineral-efficient) have a better ability than others to extract minerals from deficient soils, or a better capacity to utilise those minerals absorbed in dry matter production. In other words, the efficient strain is capable of producing more dry matter at a lower concentration of elements present in its tissues than an inefficient strain (Gerloff and Gabelman 1983).

Utilisation efficiencies for Fe, P, K and B have been reported to be governed both qualitatively and quantitatively (Table 2.2). In barley, Mn-inefficiency appears to be either under the control of a dominant gene or is linked with some related desirable traits (Blum 1988). According to Graham (1982), inheritance of Mn-efficiency is quantitative. Additive gene action has generally been found for P, K, Mg, Cu, B, Zn, Mn, Al and Fe concentrations in the ear leaf of corn (Gorsline et al. 1964b). Additive gene action was also present for P, K, Mg, Cu, Zn, Mn and Fe concentrations of the mature grain. But non-additive gene action was observed for leaf concentrations of some elements and for grain concentration of K (Gorsline et al. 1964b). Since ear leaf concentrations of various elements were not correlated with those of the grains, different genes were considered to be involved during distribution and deposition of these elements. Inheritance for Sr, Ca, Mg, K, P, Zn, Cu, B, Al, Fe and Mn accumulation in maize leaves was associated with two or three genes acting in an additive manner (Gorsline et al. 1968).

Reid (1970, 1976) reported that Al tolerance in certain barley and wheat (*Triticum aestivum* L.) varieties was controlled by a single dominant gene, but other studies in wheat have shown this characteristic to be conditioned by two or more dominant genes (Briggs and Nyachiro 1988) or by multiple genes (Lafever and Campbell 1978). In corn, Al tolerance is governed by a multiple

Table 2.2. Genes determining uptake and utilization of minerals in various crops under soil stress condition

Crop/trait	Gene(s)	Reference
Rice		
1 P absorption efficiency and P-utilization efficiency	Two major non-linked genes each controlling a single character, separately	Gunawardena and Wijeratne (1978)
2 P efficiency	Multiple genes showing epistatic gene interactions	Majumder et al. (1989)
3 Zn efficiency	Three multiple genes	Afzal et al. (1980)
Maize		
1 Fe efficiency	Single recessive gene	Bell et al. (1958)
2 P efficiency	Single dominant gene	Lyness (1936)
3 Sr toxicity tolerance	Multiple genes with additive effect	Gorsline et al. (1964a)
4 Ca toxicity tolerance	Multiple genes with additive effect	Gorsline et al. (1964a)
Soybean		
1 Fe efficiency	Single dominant gene	Weiss (1943); Läuchli (1976)
	Multiple genes but primarily controlled by a single major gene	Fehr (1982)
2 P toxicity	Single major gene	Bernard and Howell (1964)
Tomato		
1 Fe efficiency	Single dominant gene	Läuchli (1976)
	Polygenes with minor effect	Brown and Wann (1982)
2 B efficiency	Single dominant gene	Wann and Hills (1973)
3 K efficiency	Multiple genes with additive effect	Makmur et al. (1978)
4 Ca efficiency	Multiple genes with additive effect	Giordano et al. (1982)
Oat		
1 Fe efficiency	Single dominant gene	McDaniel and Brown (1982)
Drybean (*Phaseolus vulgris* L.)		
1 Fe efficiency	Multiple genes but primarily controlled by two major genes	Coyne et al. (1982)
	Two complementary dominant genes	Zaiter et al. (1987)
Snapbean (*P. vulgaris* L.)		
1 K efficiency	Single recessive gene	Shea et al. (1967)
Celery		
1 B efficiency	Single dominant gene	Pope and Munger (1953b)

Table 2.2. (*Contd.*)

Crop/trait	Gene(s)	Reference
2 Mg efficiency	Single dominant gene	Pope and Munger (1953a)
Sunflower		
1 B efficiency	Multiple genic inheritance	Blamey et al. (1984)
Triticale		
1 Cu efficiency	Single dominant gene	Graham (1982)
Barley		
1 Mn efficiency	Multiple genic inheritance	Graham (1982)

allelic series of a single gene (Rhue et al. 1978). In sorghum, Al tolerance appears to be determined by a dominant gene (Furlani et al. 1983), but other studies (Boye-Goni and Marcarian 1985) have suggested it to be a quantitative trait showing a high degree of narrow-sense heritability. In Brazillian upland rice, tolerance for Al toxicity is controlled by more than one pair of non-additive genes (Anjos et al. 1981).

Tolerance to Fe toxicity in rice was reported to be controlled by a dominant gene in one cultivar but by a recessive gene in another (Abifarin 1986): some studies have reported a complex pattern of inheritance for this trait (Senaratne unpubl; Virmani 1979). It has been hypothesised that the Fe toxicity tolerance in rice is conditioned by three genes; two act in complementary fashion while the third is an inhibitory gene (Virmani 1979).

Mn tolerance in soybean is multigenic, also showing influence of maternal effects (Brown and Devine 1980). In alfalfa, tolerance to Mn toxicity has been reported to be a quantitative trait being controlled by a number of additive genes (Dessureaux 1959). In lettuce, inheritance of tolerance to Mn toxicity varies from one to four gene loci, three of which are reported to be linked (Eenink and Garretsen 1977).

Salinity tolerance is a complex character which changes with plant age and it has, therefore, been difficult to determine patterns of inheritance. Additive as well as dominance effects have been found to be important for the inheritance of almost all the plant characters that are directly related with salt tolerance in rice (Akbar et al. 1986). At the seedling stage, shoot strength, shoot and root dry weights as well as Na and Ca content in shoots showed additive effects with high heritability. It has been demonstrated by Yeo et al. (1988) that genetic diversity for sodium uptake and hence salt tolerance persists within modern rice cultivars (IR20 and IR36). Different studies with rice have reported: three groups of genes controlling the inheritance of Na and Ca content in rice shoots (Akbar et al. 1986), two groups of genes for controlling salt tolerance (Gregorio 1991), and control of salt tolerance by genes with additive effects (Jones and Stenhouse 1984). In the latter study, the F_2 generation showed transgressive segregation with broad sense heritability ranging from 49 to 83% (Jones and Stenhouse 1984). Tolerance to salinity-induced panicle sterility in rice was dominant and controlled by at least three pairs of genes (Akbar and Yabuno 1977). According to Narayanan et al. (1990), an additive gene effect for grain yield revealed in the F_1 generation was highly significant under salt stress but non-significant in normal soil. In the F_2 generation, grain yield and seed setting percentage exhibited pronounced additive gene action, but panicle weight and number of tillers showed significant dominance gene action under saline condition.

In sorghum, Ratanadilok et al. (1978) reported salt tolerance to be controlled by complementary genes, or additive genes or genes with incomplete to complete dominance. In soybean, Cl-exclusion, a major attribute of salinity tolerance, was controlled by a dominant gene (Abel and MacKenzie 1964) but Cl-exclusion in grapevine is quantitatively inherited and showed transgressive segregation (Downton 1984). In cucumber, tolerance to salinity-induced leaf necrosis was conditioned by a dominant gene and many minor genes (Jones 1984). Narrow sense heritability for the tolerance ranged from 41 to 86%.

2.3.1 Gene Location and Linkages

Knowledge of the way genes influencing a plant's response to a soil stress are organised in the chromosome complement helps in predicting their behaviour in

Table 2.3. Gene loci determining various problem soil tolerance and their chromosome location

Crop/gene	Chromosome[a]	Reference
Wheat		
1 Al tolerance	5 D	Prestes et al. (1975)
	6 AL 7 AS 2 DL, 4 DL and 6 R	Aniol and Gustafson (1984)
2 Acidity tolerance	D	Slootmaker (1974)
Maize		
1 Accumulation of Ca and P	Both the arms of 9	Naismith et al. (1974)
2 Accumulation of Mn	9 L (distal region)	Naismith et al. (1974)
3 Alkalinity tolerance	3 L and 8 L (proximal region)	Champoux et al. (1988)
Rye		
1 Al tolerance	3 R 4 R and 6 R	Aniol and Gustafson (1984)
2 Cu efficiency	One arm of a single chromosome	Graham (1978)
Triticale		
1 Cu efficiency	5 RL	Graham (1982)

[a] Standard chromosome number and arm and genome designation.

segregating populations. For example, if a gene controlling sensitivity to a mineral stress is linked with the gene governing tolerance to another stress, the intensity of linkage between the two will determine the size of progeny population required for recovering the desired recombinants.

Geneticists have been successful in locating only few genes that determine tolerance to mineral deficiency or toxicity on specific chromosome(s) in different crops (Table 2.3). According to Mugwira et al. (1976), the ability of triticale plants to take up efficiently Ca, Mg, Al, Mn and P is inherited from the wheat genome and tolerance for soil acidity from the rye genome. The rye genes for Al tolerance are hardly expressed if they are brought together with wheat complements in the hybrid, triticale; perhaps due to their suppression by some unknown factors of the wheat genome. The A genome of *Triticum* species contributes tolerance to moderate soil acidity whereas the D genome, particularly of *Triticum aestivum*, carries one or two genes for tolerance to high soil acidity (Slootmaker 1974). A higher tolerance of soil acidity may contribute to the wider adaptability of hexaploid wheats over those of lower ploidy.

2.3.2 Correlated Changes

Selection alters the mean performance of the plant characteristics of a population by changing gene frequencies. Similarly, introgression of new genetic traits also brings about changes in the average performance of a breeding line. Pleiotropism, genetic linkage and coherence (i.e. non-random segregation) are the causal factors for inducing such changes.

The kernel protein content in wheat has been reported to be positively associated with tolerance to soil acidity (Mesdag et al. 1970), indicating possible linkage between the two characteristics. In maize, the gene for heat tolerance also controls Al tolerance, which is an example of pleiotropism (de Miranda et al.1984). Correlated changes have also been observed in the case of plants grown on B- or Fe-deficient soils. In tomato, the recessive gene *btl* controls brittle leaf and stem characteristics but it also conditions the inefficient transport of B from the roots to shoot (Wall and Andrus 1962). Fe-efficient tomato genotypes have greater tolerance to heavy metals (Brown and Jones 1975) and have higher nitrate reductase activity (Brown and Jones 1976). It has also been observed that Fe-inefficient tomato plants develop lateral roots in Fe-deficient conditions (Brown and Ambler 1974). All these findings, therefore, indicate pleiotropic control of correlated changes. Fe-inefficient plants also accumulate higher concentrations of interacting elements (Ca and P). For example, an Fe-inefficient oat contained twice as much Ca as an Fe-efficient cultivar (Brown and McDaniel 1978) and a sorghum genotype tolerant to low soil pH contained twice as much P as a sensitive genotype (Brown et al. 1977). In maize and soybean, Fe-inefficient genotypes possessed high levels of P, Ca, Zn and Mn, showing interaction with Fe (Olsen 1972). Cu-efficiency in triticale is linked with the gene for hairy peduncles (Graham 1982).

2.4 Crop-Improvement

Conventional methods of breeding that have been used to identify/develop crop varieties tolerant to soil stresses are: (1) introduction, (2) pure line and mass selection, (3) hybridisation and selection, (4) mutation breeding, (5) polyploid breeding and (6) heterosis breeding.

2.4.1 Introduction

The introduction of plants to new areas can accomplish results similar to those otherwise achieved by developing varieties through hybridisation and selection. Mankind's mobility from place to place has helped the spread of plant materials around the world and, today as in the past, introductions play a major role in varietal improvement. The majority of traditional varieties grown on problem soils of South and Southeast Asia appear to have originated from India and Thailand. The importance of plant introductions has recently been recognised and plant material exchanged at both national and international level, although most introductions are not adequately recorded. At present, there are 113 organisations in 53 different countries and 12 international centres (Hanson et al. 1984) involved in transferring genetic material between regions.

International Agriculture Research Centres (IARCs) have established mechanisms for the exchange of improved germplasm. These can also be considered as introductions. For example, the International Network for Genetic Evaluation of Rice (INGER) distributes to all interested agencies improved rice germplasm developed at IRRI and in national programmes. Among the 25 different nurseries it assembles and distributes every year, three are for problem soil environments. Germplasm for salinity, alkalinity and acidity are available in these nurseries. Similar programmes are in operation in the International Institute of Tropical Agriculture (IITA), the Centro Internacional de Agricultura Tropical (CIAT), the International Crop Research Institute for the Semi-Arid Tropics (ICRISAT), and the Centro Internacional de Mejoramiento de Maiz y Trigo (CIMMYT).

2.4.2 Pure Line and Mass Selection

The full spectrum of genetic variability for tolerance to problem soils available in a crop should be first screened and evaluated before embarking on hybridisation programmes. Landraces of self-pollinated crops usually consist of a heterogeneous mixture of many, highly homozygous genotypes. It is, therefore, possible to select within a landrace for a better individual, the progeny of which will be genetically identical. The procedure of pure line breeding involves initial selection of a large number of individual plants (pure lines) from a genetically diverse landrace. Thereafter, their progeny performances are visually evaluated. The lines with obvious defects are discarded. The remaining lines are tested in replicated trials and the best selected as a variety. Tolerant strains, or those possessing higher efficiency for acquisition, transport and utilisation of essential soil elements may be selected from landraces collected from areas where the stresses occur. In rice, out of a considerable number of landraces collected from different salt-affected ricelands of India, pure line selection has led to the development of 18 salt-tolerant varieties (Table 2.4) some of which have been introduced to other countries as varieties or donors for salt tolerance (IRRI 1984; Sinha and Bandyopadhyay 1984). In the grass *Agrostis tenuis*, ecotypes tolerant to four heavy metals (Cu, Ni, Zn and Pb) were collected by the University of Liverpool, England, out of which those identified as tolerant to specific minerals were released as cultivars (Humphreys and Bradshaw 1976). At CIAT, a number of high-yielding cassava cultivars tolerant to Al-toxic soils, has been selected from a collection of pure lines (CIAT 1977).

The basic difference between pure line and mass selection is whether one or a number of selected lines forms the new variety. In pure line breeding, the new variety is derived from the progeny of a single pure line whereas in mass selection, from the progeny of many pure lines. The procedure of mass selection involves selection of a larger number of desirable plants from a population of a heterogeneous mixture of genotypes followed by harvesting their seed in bulk without any progeny test. The selected material can be planted and the selection

Table 2.4. Pure line selection from local materials of India

Selection identity	Parent material	Region of adaptation
Jhona 349	Jhona	Punjab and Haryana
Patnai 23	Patnai	West Bengal
CSR-1	Damodar	West Bengal
CSR-3	Getu	West Bengal
CSR-2	Dasal	West Bengal
Matla	Beni sail	West Bengal
Hamilton	Nona Bokra	West Bengal
SR 26B	Kalambank	Wide Adaptability
KR 1-24	Kala Ratta	Maharashtra
BR4-10	Bhura Ratta	Maharashtra
MCM-2	Buddamolagolakulu	Andhra Pradesh
Arya 33	Arya	Karnataka
Mo 1	Chettivirippu	Kerala
Mo 2	Kalladachampavu	Kerala
Mo 3	Kunjathikkara	Kerala
Vytilla 1	Choottupokkali	Kerala
CSR-6	Nonasail	Orissa and W Bengal

Source: adapted from Proceedings of an IRTP/INSFFER Monitoring Program. Rice Improvement for Adverse Soils: with emphasis on acid sulfate and coastal salinity, IRRI (1984).

repeated in the following generation of selection. Therefore it is possible, over generations, to increase the frequency of desirable genotypes without upsetting the agronomic features of the population.

Mass selection for salt tolerance in wheat has been performed under salinized irrigation water. Selection was initially made at a salt concentration of 6000 ppm followed by higher concentrations, up to 12 000 ppm, causing an elimination of 50% of the population (Dewey 1962). Five cycles of mass selection for salinity tolerance of germinating seeds have been conducted and found to be effective in alfalfa (Allen et al. 1985). Germination under stress increased substantially by the end of the final cycle of selection. Three cycles of mass selection have been applied for Al tolerance on four heterogeneous populations of grain *Amaranthus* (of the species *A. hypochondriacus*, *A. cruentus* and *A. hybridus*: Campbell and Foy 1987). For the first two cycles, populations were screened on acid soil for Al tolerance only. Selected plants were selfed and evaluated for superior agronomic characters in the field. Thereafter, seeds of agronomically vigorous plants were grown in a growth chamber for selecting Al tolerant plants in a third cycle followed by another field evaluation for general vigour. Selection for Al tolerance was found to be effective only in the *A. cruentus* population.

2.4.3 Hybridisation and Selection

An essential role of the plant breeder is to create variability by way of making crosses between varieties of known performance. Selection for desirable recombination in segregating generations is practised in many ways.

2.4.3.1 Pedigree Method

In this method, desirable plants are selected in F_2 to establish F_3 generation pedigrees. In F_3, plant selections are made from promising F_3 lines. The procedure is continued in all subsequent generations of inbreeding, until homozygous lines are produced. Therefore, selection in a segregating population is based on the recorded performance from parents to progenies.

The pedigree method has been used in IRRI to transfer salt tolerance of traditional rice cultivars to modern high yielding types, although the level of tolerance of new derivatives is lower than that of traditional cultivars. Promising lines (and their tolerance donors) are IR4595-4-1-13 (Pokkali), IR9884-54-3 and IR10198-66-2 (Nona Bokra) and IR10206-29-2-1 (SR26B). Some salt-tolerant rice varieties that have been developed elsewhere by the pedigree method are given in Table 2.5.

2.4.3.2 Bulk Method

In this method, individuals of a segregating population of a cross are grown en masse with the aim of allowing natural selection to weed out weak plants. The

Table 2.5. List of salt-tolerant rice varieties developed by the pedigree method

Variety developed	Parentage	Reference
PVR-1	SR 26B/MTU-1	Rana (1986)
MCM-1	Co 18/Kuthir	Rana (1986)
SR 10022	SR 26B/MTU-1	Rana (1986)
SR 1-2-1	Jaya/SR 10022	Rana (1986)
MR-18	SR 26B/Wannar-1	Rana (1986)
Co 43	Dasal/IR20	Rana (1986)
Usar 1	Jaya/Getu	Rana (1986)
MK 47-22	Malkudai/DR 1-24	Salvi and Chavan (1983)
SR 3-9	KR1-24/Zinnya 149	Salvi and Chavan (1983)
Panvel 28-23	Blue Belle/BR 4-10	Salvi and Chavan (1983)
Panvel 11-2	TNI/BR 4-10	Salvi and Chavan (1983)
Panvel 5-30 (Panvel-1)	IR 8/BR4-10	Salvi and Chavan (1983)
Panvel 32-10-1-1	BR 4-10/IR8	Salvi and Chavan (1983)
Yeonggwang	Gudaenaejouk 3/Eunbangju	YT Lee (pers. comm.)
Gancheok 9	Gudaenaejouk 3/Eunbangju	YT Lee (pers. comm.)
Namyangbyeo	Fuji 280/BL1	YT Lee (pers. comm.)
Seohaebyeo	CS-SR/Fuji 280	YT Lee (pers. comm.)
Janganbyeo	Inabawase/Dongjinbyeo	YT Lee (pers. comm.)
Seoanbyeo	Suweon 224/Inabawase//	YT Lee (pers. comm.)
	Seolagbyeo	YT Lee (pers. comm.)
Gyehwabyeo	Iri 348/Saikai 145	YT Lee (pers. comm.)
Yeongdeogbyeo	Milyang 15///Pebihon/Kanto 98//	YT Lee (pers. comm.)
	Kanto 100	YT Lee (pers. comm.)
Donghaebyeo	Milyang 20/Milyang 15	YT Lee (pers. comm.)

survivors are bulked to form the next generation. The procedure is repeated until an agronomically uniform bulk is produced.

Soil stresses vary widely both over time and space, more so in natural than in experimental conditions. As a consequence, population size would need to be increased and evaluations would have to be repeated in field conditions. In such a situation, the pedigree method becomes impractical and the most suitable substitute is the "bulk method". The classical bulk method is rarely used, but instead a "modified bulk method" is employed in which only desirable genotypes or individuals are selected and bulked. In principle, the segregating materials are evaluated and selected under stress conditions for several successive generations as bulks. When the desired level of tolerance is achieved, plant selections are made and the pedigree method is followed for their further selection and purification. Final testing of the selected material is performed in the target environment. In practice, however, breeders vary this procedure according to their resources and other constraints, as illustrated in the following examples.

In Fiji, a cross was made between a traditional rice variety, Lalka Motka (tall, adapted to acid sulphate, saline and peat soil conditions, and photoperiod-sensitive) and an IRRI line IR661 (semi-dwarf, photoperiod-insensitive). Segregating material was grown as a bulk from F_2 to F_4 and bulk-selected for semi-dwarf and photoperiod-insensitive plants. In the F_5 generation, progenies were selected and grown in pedigree rows in a multiple stress environment. The promising lines were selected and tested at a number of locations in different years and one line, K127-20-1, which significantly out-yielded the standard check in acid soil areas was released as Deepak (Reddy et al. 1987).

In Sri Lanka, out of three elite rice lines developed for tolerance to salinity from a cross made between BG94-1 and Pokkali, an At 69-2 line has been released in 1989 for cultivation in coastal saline areas (S. Abeysekara pers. comm.). The segregating population was grown as bulk up to F_4 on normal soil. Selection was based on plant type, grain yield, and disease and pest resistance. In F_4, individual plant selections were made. In subsequent generations plant selections were evaluated by the pedigree method under naturally occurring saline soil conditions.

Jones (1989) has attempted to improve the salt tolerance of photoperiod-sensitive mangrove swamp rice varieties. F_2 seeds of crosses were sown in plastic basins containing mangrove soil. After 4 weeks, plants were grown in saline culture solution for 2 weeks. Surviving individuals were subjected to short-day treatment for 2 weeks to shorten the maturation period. Seeds were harvested in bulk and the cycle was repeated for three more generations (F_3–F_5). It was observed that the number of surviving plants and filled grains increased with each generation of selection. Advanced families were, thereafter, grown in mildly saline mangrove swamps for further selection.

Fe-tolerant improved rice varieties have been developed in Sri Lanka by a modified bulk method in which materials of all generations, including F_1, have been evaluated and selected under naturally occurring Fe-toxic soil conditions (P. E. Peiris pers. comm.).

In a wheat breeding programme at Londrina, Parana, in Brazil, individual plants were selected in F_2 and F_5, but were bulk-selected in F_3 and F_4 generations grown on Al-toxic and P-deficient soils (Reide and Campos 1988).

IRRI's current breeding programme for problem soils uses the modified bulk method. In the F_2 generation, materials are evaluated and bulk selected under naturally occurring stress conditions at experimental sites in the Philippines having different types of acid, saline and acid-sulphate soils. At F_3, materials are evaluated on normal soil to select for agronomic traits such as plant type, lodging resistance, grain quality, and pest and disease resistance. Shuttling between stress and normal soil proceeds up to about F_6, when plant selections are made. Bulk harvests of seed from good pedigree lines are given to collaborators as advanced bulks for evaluation and selection in target environments. In this breeding programme, shuttling between stress and normal soil is done because the problem soil could be cultivated only once a year (during the rainy season).

2.4.3.3 Backcross Breeding

When the objective is to transfer one or two major genes from one cultivar (donor parent) to another (recurrent parent), the backcross method of breeding is used. The initial single cross between the two parents is repeatedly backcrossed to the recurrent parent so that individuals having all the traits of the recurrent parent plus the major gene(s) of the donor parent can be recovered during selection.

In Brazil, backcross breeding has been initiated to transfer the P-efficiency characteristic of two wheat cultivars to high yielding wheat varieties (Rosa 1988). Plants in segregating generations have been selected under low pH and low soil-P. Efforts are also being made to transfer genes for P-efficiency from rye to wheat by the backcross method (Rosa 1988).

An inbred-backcross method of Bliss (1981) has been used to transfer P-efficiency from an exotic common bean (*Phaseolus vulgaris* L.) to an agriculturally useful cultivar named Sanilac (Schettini et al. 1987). The method involved two successive backcrosses to Sanilac followed by three or four generations of selfing to produce advanced lines with the desired level of homozygosity. The selection among individuals was neither made in backcrosses nor in selfed progenies but was begun among advanced lines and about 10% of the lines had greater seed yield than Sanilac under the low-P soil conditions.

In tomato, attempts have been made to incorporate the salinity of a wild species, *Lycopersicon cheesmanii*, into the cultivated species, *L. esculentum*, by the backcross method of breeding (Rush and Epstein 1981).

A backcross as well as a single cross have been used in an attempt to transfer the Fe-efficiency of two soybean lines to varieties which are high yielding but otherwise sensitive to Fe deficiency (Hintz et al. 1987). The backcross populations (derived from one backcross followed with three generations of selfing) had higher seed yield but lower tolerance than the single cross populations. However, the difference in yield was significant in only one of the four populations. On the other hand, three of the four single cross populations were significantly superior in tolerance to the backcross populations. Since the handling of single cross populations requires less labour and time, it is considered a better method than backcorssing.

2.4.3.4 Recurrent Selection

This is broadly defined as the methodical selection of promising plants from a population and their inter se recombination to form a new population for reselection. In crops, adaptability and productivity are mostly negatively correlated. Therefore, recurrent selection is employed to break undesirable linkages and to increase the frequency of favourable combinations, particularly where the character concerned is influenced by genes with additive effects. The methodology of recurrent selection has been applied for increasing tolerance to Fe-deficiency chlorosis in soybean (Prohaska and Fehr 1981).

In alfalfa, recurrent selection has been effective in raising tolerance to an acid and Al-toxic soil. A broad-base germplasm of alfalfa was subjected to two cycles of divergent selection in acid soil for tolerance, as well as for susceptibility to Al. Finally, Al-tolerant and susceptible derivatives were tested and were found to be significantly different from each other (Devine et al. 1976).

2.4.3.5 Rapid Generation Advance Procedures

Most problem soils are rainfed, and cultivation in such areas is restricted to the rainy season only. This slows down breeding progress since only one generation of the crop can be raised each year. Shuttle breeding and the single seed descent method are being employed to overcome this difficulty.

In 1974, Borlaug first launched a cooperative research programme in wheat between CIMMYT and research institutes of Brazil. In this programme, successive generations were alternated between CIMMYT and Brazil and selections were continued until true breeding lines were produced. It has been possible to combine genes for high yield from Mexican wheats with those conferring the Al-tolerance of Brazilian wheats through this "shuttle breeding", as a result of which several cultivars have been developed and recommended for cultivation on acid soils of Brazil (Rajaram et al. 1987).

The single seed descent method (SSD), also known as the modified pedigree method (Johnson and Bernard 1962; Brim 1966), has proven itself to be very useful in rapidly advancing the generations of segregating material. In this method, segregating materials are rapidly inbred while selection remains suspended. At each generation, one seed harvested from each plant is bulked to establish the next generation. The procedure is repeated as necessary and the advanced bulk can be evaluated normally to select desirable individuals. Since only a single seed is required as the harvest of an individual, segregating populations can be planted at very high density. Therefore in SSD, a large number of populations can be handled in a relatively small space. Moreover, at high densities, plants flower early, and as a result, generation time is reduced considerably.

In rice, photoperiod-sensitivity is often essential for some problem soil lands and a special facility called the rapid generation advance facility (RGA) has been constructed at IRRI to handle photoperiod-sensitive breeding materials (Ikehashi and HilleRisLambers 1979; Pateña et al. 1980; Vergara et al. 1980). In this technique, segregating populations are grown at very close spacing (about 1000 plants/m^2) under relatively high temperatures and short days. About three generations can be advanced within a year. The usual practice is to evaluate an F_2 generation during the cropping season, bulk harvest desirable genotypes and advance the next generation or two in the RGA facility without any selection. The harvest from the RGA is established in the field for further selection. The facility is also used in IRRI in the same manner during the dry season for photoperiod-insensitive breeding materials, where field evaluation is possible only in the wet season. According to Jones (1989), four generations of photoperiod-sensitive mangrove swamp rice varieties can be raised in a year simply by shortening the day length for 2 weeks after 6 weeks of growth.

2.4.4 Mutation Breeding

If genetic variability for tolerance to a mineral stress is inadequate or not available, then mutagenesis is one way to induce variability. Mutation breeding has rarely been applied for developing tolerance to problem soils, possibly because existing genetic variability has not yet been fully explored.

Fertilized egg cells of the salt-sensitive rice variety Taichung-65 were treated with different doses of N-methyl N-nitrosourea. In the third generation (M_3), two salt-resistant mutants were detected with increased survival with respect to the original parent (IRRI 1984). Chemical mutagens have been used to induce dwarfism into traditional tidal swamp rice varieties of Indonesia (Mahadevappa et al. 1981). Irradiated seeds of Basmati 370 rice produced a salt-tolerant mutant, RST-24 (Sajjad 1990). In India, two promising rice mutants; Mut 1 (CSR-4) and Mut 2 have been developed for salt tolerance from the cultivar IR-8 (Sinha and Bandyopadhyay 1984). The mutant variety CSR-4 has also been outstanding in saline alkali soils of Northern India (Yadav 1979).

A Mn-efficient barley mutant has been induced by irradiation (Anonymous 1984). The mutant has performed better and also had a higher concentration of Mn in plant tissues than the source material when grown on Mn-deficient soils. Six salt-tolerant mutants of *Arabidopsis thaliana* that required high osmotic pressure for normal growth were developed through mutation breeding (Langridge 1958).

2.4.5 Polyploid Breeding

If variability for a desired character is restricted in any crop, it may sometimes be increased through reduplication of chromosome sets (i.e. induction of polyploidy). It has been observed that allopolyploid crop species are more tolerant to both alkali and saline soil conditions than their diploid counterparts (Rana 1978; Rana et al. 1980). Hexaploid wheats have higher tolerance to alkali and saline soil conditions than their tetraploid and diploid counterparts. Diploid einkorn wheats do not even flower in alkali or saline soils. With increasing ploidy level, the efficiency of sulphate and of K uptake increased in wheat and sugar beet, but not in tomato (Cacco et al. 1976). Similarly, *Brassica juncea* (Indian mustard), an allopolyploid species, is found superior in salt tolerance to its diploid progenitors, *B. campestris* and *B. nigra*. An exception is found in cotton where the diploid Asiatic species, *Gossypium arboreum* and *G. herbaceum*, are more tolerant than both allopolyploid species, *G. hirsutum* and *G. barbadense* (Rana 1986).

2.4.6 Heterosis Breeding

Hybrid cultivars, particularly of cross-pollinated crops, have commercial importance, as heterozygosity tends to hide undesirable effects of recessive genes as well as expressing desirable effects of dominant genes. Most tolerance traits are dominant and reveal heterosis in F_1 hybrids. It is also possible to bring together in F_1 hybrids tolerance to one major stress from one parent with that to another stress from the another parent.

In rice, hybrids have revealed superior fertiliser use and root efficiency (McDonald et al. 1971; Lin and Yuan 1980; Ekanayake et al. 1986) and enhanced tolerance to adverse soil conditions (Akbar and Yabuno 1975; Senadhira and Virmani 1987; Moeljopawiro and Ikehashi 1981).

In sorghum, when a male-sterile acid-susceptible line was crossed to an acid-tolerant fertility restorer, it produced an F_1 hybrid that gave enhanced grain yield on both acid and non-acid soils (Humberto et al. 1988). Cacco et al. (1978) have reported heterosis for sulphate uptake in maize hybrids.

2.5 Summary

Both inter- and intra-specific variability for tolerance to mineral stresses exist in crop species. A wide range of methods is potentially available to exploit such variation in the development of new cultivars, but to date conventional plant breeding methods are the only ones to have been used widely.

References

Abel GH, MacKenzie AJ (1964) Salt tolerance of soybean varieties (*Glycine max* L Merrill) during germination and later growth. Crop Sci 4: 157–161

Abifarin AO (1986) Inheritance of tolerance to iron toxicity in rice cultivars. In: Rice genetics. Proc Intl Rice Genetics Symp 27–31 May 1985. IRRI, Manila, Philippines, pp 423–427

Afzal M, Ahmad M, Ali M, Yousuf M (1980) Studies on the inheritance of resistance to zinc deficiency in rice (*Oryza sativa* L.) J Agric Res 18: 49–54

Akbar M, Yabuno T (1974) Breeding for saline-resistant varieties of rice. II, Comparative performance of some rice varieties to salinity during early development stage. Jpn J Breed 24: 176–181

Akbar M, Yabuno T (1975) Breeding for saline-resistant varieties of rice III. Response of F_1 hybrids to salinity in reciprocal crosses between Jhona 349 and Magnolia. Jpn J Breed 25: 215–220

Akbar M, Yabuno T (1977) Breeding for saline-resistant varieties of rice. IV Inheritance of delayed-type panicle sterility induced by salinity. Jpn J Breed 27: 237–240

Akbar M, Khush GS, HilleRisLambers D (1986) Genetics of salt tolerance in rice In: Rice Genetics Proc Int Rice Genetics Symp 27–31 May 1985. IRRI, Manila, Philippines, pp 399–409

Allen SG, Dobrenz AK, Schonhorst M, Stoner JE (1985) Heritability of NaCl tolerance in germinating alfalfa seed. Agron J 77: 99–101

Aniol A, Gustafson JP (1984) Chromosome location of genes controlling aluminum tolerance in wheat rye and triticale. Can J Genet Cytol 26: 701
Anjos CV, Nguyen TV, Silva JC, Galvao JO (1981) Inheritance of tolerance for aluminum toxicity in Brazilian rice. Int Rice Res Newsl 6(4): 9
Anonymous (1984) Performance of a mutant barely on Mn-deficient soil. Mutat Bredding Newsl 24: 18
Antonovics J, Bradshaw AD, Turner RG (1971) Heavy metal tolerance in plants. Adv Ecol Res 7: 1–85
Armiger WH, Foy CD, Flaming AL, Caldwell BE (1968) Differential tolerance of soybean varieties to an acid soil high in exchangeable aluminum. Agron J 60: 67–70
Aslam M (1989) Status of salinity research in Pakistan Working Group Meeting on Collaborative Research on the Development of Salinity Tolerant Rice Cultivars. May 11–13 1989. IRRI, Los Banos, Philippines
Barakat MA, Khalid MM, Atia MH (1971) Effect of salinity on the germination of 17 rice varieties. Agric Res Rev UAR 49: 219–224
Bari G, Hamid A (1988) Salt tolerance of rice varieties and mutant strains. Pak J Sci Ind Res 31: 282–284
Bell WD, Bogorad L, McIlrath WJ (1958) Response of the yellow-stripe maize mutant (ys1) to ferrous and ferric iron. Bot Gaz 120: 36–39
Bernard RL, Howell RW (1964) Inheritance of phosphorus sensitivity in soybeans. Crop Sci 4: 298–299
Blamey FPC, Vermeulen WJ, Chapman J (1984) Inheritance of boron status in sunflower. Crop Sci 24: 43–46
Bliss FA (1981) Utilization of vegetable germplasm. HortScience 16: 129–132
Blum A (1988) Plant breeding for stress environments CRC, Boca Raton
Boye-Goni SR, Marcarian V (1985) Diallel analysis of aluminum tolerance in selected lines of grain sorghum. Crop Sci 25: 749–752
Briggs KG, Nyachiro JM (1988) Genetic variation for aluminum tolerance in Kenyan wheat cultivars. Commun Soil Sci Plant Anal 19: 1273–1284
Brim CA (1966) A modified pedigree method of selection in soybeans. Crop Sci 6: 220
Brown JC, Ambler JE (1974) Iron-stress response in tomato (*Lycopersicon esculentum*) 1 Sites of Fe reduction absorption and transport. Physiol Plant 31: 221–224
Brown JC, Devine TE (1980) Inheritance of tolerance or resistance to manganese toxicity in soybeans. Agron J 72: 898–904
Brown JC, Jones WE (1975) Heavy-metal toxicity in plants IA crisis in embryo. Commun Soil Sci Plant Anal 6: 421–438
Brown JC, Jones WE (1976) Nitrate reductase activity in calcifugous and calcicolous tomatoes as affected by iron stress. Physiol Plant 38: 273–277
Brown JC, Jones WE (1977a) Fitting plants nutritionally to soils I. Soybeans. Agron J 69: 399–404
Brown JC, Jones WE (1977b) Fitting plants nutritionally to soils II. Cotton. Agron J 69: 405–409
Brown JC, Jones WE (1977c) Fitting plants nutritionally to soils III. Sorghum, Agron J 69: 410–414
Brown JC, McDaniel ME (1978) Factors associated with differential response of oat cultivar to iron stress. Crop Sci 18: 551–556
Brown JC, Wann EV (1982) Breeding for iron efficiency: use of indicator plants. J Plant Sci Nutr 5: 623–635
Brown JC, Ambler JE, Chaney RL, Foy CD (1972) Differential responses of plant genotypes to micronutrients. In: Mortvedt JJ, Giordano PM, Lindsay WL (eds) Micronutrients in agriculture. Soil Science Society of America, Madison, WI
Brown JC, Clark RB, Jones WE (1977) Efficient and inefficient use of phosphorous by sorghum. Soil Sci Soc Am J 41: 747–750
Cacco G, Ferrari G, Lucci G (1976) Uptake efficiency of roots in plants at different ploidy levels. J. Agric Sci 87: 585–589
Cacco G, Ferrari G, Saccomani M (1978) Variability and inheritance of sulphate uptake efficiency and ATP-sulfurylase activity in maize. Crop Sci 18: 503–505

Campbell TA, Foy CD (1987) Selection of grain *Amaranthus* species for tolerance to excess aluminum in an acid soil. J Plant Nutr 10: 249–260
Carlson JR Jr., Ditterline RL, Martin JM, Sands DC, Lund RE (1983) Alfalfa seed germination in antibiotic agar containing NaCl. Crop Sci 23: 882–885
Carter OG, Rose IA, Reading PF (1975) Variation in susceptibility to manganese toxicity in 30 soybean genotypes. Crop Sci 15: 730–732
Champoux MC, Nordquist PT, Compton WA, Morris MR (1988) Use of chromosomal translocations to locate genes in maize for resistance to high pH soil. J Plant Nutr 11: 783–791
CIAT (Centro International de Agricultura Tropical) (1977) Annu Rep Cali Colombia
Clark RB, Brown JC (1980) Role of the plant in mineral nutrition as related to breeding and genetics In: Murphy LS, Doll EC, Welch LF (eds) Moving up the yield plateau: advances and obstacles. Soil Science Society of America, Madison, WI pp 45–70
Coronel VP, Akita S, Yoshida S (1990) Aluminum toxicity tolerance in rice (*Oryza sativa*) seedlings. In: van Beusichen ML (ed) Plant nutrition – physiology and applications. Kluwer Academic, Dordrecht, pp 357–363
Coyne DP, Korban SS, Knudsen D, Clark RB (1982) Inheritance of iron deficiency in crosses of dry beans (*Phaseolus vulgaris* L.). J Plant Nutr 5: 575–585
Datta SK (1972) A study of salt tolerance of twelve varieties of rice. Curr Sci 41: 456–457
de Miranda LT, de Miranda LEC, Sawazaki E (1984) Genetics of environmental resistance and super-genes: latent heat tolerance. Maize Genet Coop Newsl 58: 48–50
Dessureaux L (1959) Heritability of tolerance to manganese toxicity in lucerne Euphytica 8: 260–265
Dessureaux L, Ouellette GI (1958) Chemical composition of alfalfa as related to the degree of tolerance to manganese and aluminum. Can J Soil Sci 38: 206–214
Devine TE, Foy CD, Fleming AL, Hanson CH, Campbell TA, McMurtrey JE III, Schwartz JW (1976) Development of alfalfa strains with differential tolerance to aluminum toxicity. Plant Soil 44: 73–79
Dewey DR (1962) Breeding crested wheatgrass for salt tolerance. Crop Sci 2: 403–407
Downton WJS (1984) Salt tolerance of food crops: prospectives for improvement. CRC Crit Rev Plant Sci 1: 183
Eenink AH, Garretsen F (1977) Inheritance of insensitivity of lettuce to a surplus of exchangeable manganese in steam-sterilized soils. Euphytica 26: 47–53
Ekanayake IJ, Garrity DP, Virmani SS (1986) Heterosis for root pulling resistance in F_1 rice hybrids. Int Rice Res Newsl 11(3): 6
Epstein E (1972) Mineral nutrition of plants: principles and perspectives. Wiley, New York
Fehr WR (1987) Mass selection in self pollinated populations. In: Principles of cultivar development, vol I. Theory and technique. McGraw, New York, pp 328–331
Foy CD, Chaney RL, White MC (1978) The physiology of metal toxicity in plants. Annu Rev Plant Physiol 29: 511–566
Foy CD, Oakes AJ, Schwartz JW (1979) Adaptation of some introduced *Eragrostis* species to calcareous soil and acid mine spoil. Commun Soil Sci Plant Anal 10: 953–968
Francois LE, Bernstein L (1964) Salt tolerance of safflower. Agron J 56: 38–40
Furlani PR, Clark RB, Ross WM, Maranville JW (1983) Variability and genetic control of aluminum tolerance in sorghum genotypes. In: Saric MR, Loughman BC (eds) Genetic aspects of plant nutrition. Nijhoff, The Hague, pp 453–461
Gerloff GC, Gabelman WH (1983) Genetic basis of inorganic plant nutrition. In: Encyclopedia of plant physiology, vol 15. Inorganic plant nutrition. Springer, Berlin Heidelberg, New York pp 454
Gildersleeve RR, Ocumpaugh WR (1988) Variation among *Trifolium* species for resistance to iron-deficiency chlorosis. J Plant Nutr 11: 727–737
Giordano L, de B Gabelman WH, Gerloff GC (1982) Inheritance of differences in calcium utilization by tomatoes under low-calcium-stress. J Am Soc Hortic Sci 107: 664–669
Gorsline GW, Thomas WI, Baker DE, Ragland JL (1964a) Relationship of strontium–calcium accumulation within corn. Crop Sci 4: 154–156

Gorsline GW, Thomas WI, Baker DE (1964b) Inheritance of P, K, Mg, Ca, B, Mn, Al and Fe concentration by corn (*Zea mays* L.) leaves and grain. Crop Sci 4: 207–210
Gorsline GW, Thomas WI, Baker DE (1968) Major gene inheritance of Sr, Ca, Mg, K, P, Zn, Cu, B, Al, Fe and Mn concentration in corn (*Zea mays* L.). PA State Univ Agric Exp Stn Bull 746
Graham RD (1978) Plant breeding for nutritional objectives. In: Ferguson AR, Bieleski RL, Ferguson IB (eds) Plant nutrition. Proc 8th Int Colloq Plant Anal Fert Prob Auckland, pp 165–170
Graham RD (1982) Breeding for nutritional characteristics in cereals. Adv Plant Nutr 1: 57–102
Gregorio GB (1991) Genetic components of salinity tolerance in rice (*Oryza sativa* L.). A MSc Thesis, University of Philippines
Gunawardena SDIE, Wijeratne HMS (1978) Screening of rice under phosphorus deficient soil in Sri Lanka. Int Rice Res Conf, IRRI, Manila Philippines, p 11
Hanson J, Williams JT, Freund R (1984) Institutes conserving crop germplasm: the IBPGR global network of gene banks. International Board of Plant Genetic Resources. FAO, Rome pp 25
Hill J, Robson AD, Loneragan JF (1978) The effects of copper and nitrogen supply on the retranslocation of copper in four cultivars of wheat. Aust J Agric Res 29: 925–939
Hintz RW, Fehr WR, Cianzio SR (1987) Population development for the selection of high yielding soybean cultivars with resistance to iron-deficiency chlorosis. Crop Sci 27: 707–710
Howeler RH, Cadavid LF (1976) Screening of rice cultivars for tolerance to Al-toxicity in nutrient solutions as compared with a field screening method. Agron J 68: 551–555
Humberto CC, Catalino IF, Henry VA (1988) Evaluacion de la tolerancia do sorgo hybrido (*Sorghum bicolor* L.) Moench en suelos acidos. Acta Agron 38: 23–30
Humphreys MO, Bradshaw AD (1976) Genetic potentials for solving problems of soil mineral stress heavy metal toxicities. pp 95–109. In: Wright MJ (ed) Plant adaptation to mineral stress in problem soils. Cornell University Agricultural Experiment Station, Ithaca, NY
Hutton EM, Williams WT, Andrew CS (1978) Differential tolerance to manganese is introduced and bred lines of *Macroptilium atropurpureum*. Aust J Agric Res 29: 67–79
Ikehashi H, HilleRislambers D (1979) Integrated international collaborative program in the use of rapid generation advance in rice. In: proc 1978 Int Deepwater Rice Workshop, IRRI, Philippines, pp 261–276
Ikehashi H, Ponnamperuma FN (1978) Varietal tolerance to rice for adverse soil. In: Soils and Rice. IRRI, Manila, Philippines, pp 801–823
IRRI (International Rice Research Institute) (1968) Annu Rep for 1968. IRRI, Manila, Philippines, 402 pp
IRRI (International Rice Research Institute) (1970) Annu Rep for 1969. IRRI, Manila, Philippines, 266 pp
IRRI (International Rice Research Institute) (1971) Annu Rep for 1970. IRRI, Manila, Philippines, 265 pp
IRRI (International Rice Research Institute) (1972) Annu Rep for 1971. IRRI, Manila, Philippines, 238 pp
IRRI (International Rice Research Institute) (1973) Annu Rep for 1972. IRRI, Manila, Philippines, 246 pp
IRRI (International Rice Research Institute) (1974) Annu Rep for 1973. IRRI, Manila, Philippines, 266 pp
IRRI (International Rice Research Institute) (1975) Annu Rep for 1974. IRRI, Manila, Philippines, 384 pp
IRRI (International Rice Research Institute) (1976) Annu Rep for 1975. IRRI, Manila, Philippines, 479 pp
IRRI (International Rice Research Institute) (1977) Annu Rep for 1976. IRRI, Manila, Philippines, 418 pp
IRRI (International Rice Research Institute) (1978) Annu Rep for 1977. IRRI, Manila, Philippines, 548 pp

IRRI (International Rice Research Institute) (1984) Annu Rep for 1983. IRRI, Manila, Philippines, 493 pp
IRRI (International Rice Research Institute) (1989) Annu Rep for 1988. IRRI, Manila, Philippines, 646 pp
IRTP (International Rice Testing Programme) (1988) Standard evaluation system for rice, 3rd edn. International Rice Research Institute, Manila, Philippines, 123 pp
Janardhan KV, Murty KS (1972) Studies on salt tolerance in rice III. Relative salt tolerance of some local and high yielding rice varieties. Oryza 9:23–34
Jayawardena SDG, Watabe T, Tanaka K (1977) Relation between root oxidizing power and resistance to iron toxicity in rice. Rep Soc Crop Sci Breed Kinki Jpn 22: 38–47
Johnson HW, Bernard RL (1962) Soybean genetics and breeding. Adv Agron 14: 149–221
Jones MP (1989) Rapid generation and improvement of salt tolerance in photoperiod sensitive mangrove swamp rice varieties and breeding lines. Environ Exp Bot 29: 417–422
Jones MP, Stenhouse JW (1984) Inheritance of salt tolerance in mangrove rice. Int Rice Res Newsl 9(3): 9
Jones RW Jr (1984) Studies related to genetic salt tolerance in the cucumber. Diss Abstr Int B 45: 1376
Jung GA (ed) (1978) Crop tolerance to sub-optimal land conditions. American Society of Agronomy, Madison, WI
Katyal JC, Seshu DV, Shastry SVS, Freeman WH (1975) Varietal tolerance to low phosphorus conditions. Curr Sci 44: 238–240
Lafever HN, Campbell LG (1978) Inheritance of aluminum tolerance in wheat. Can J Genet Cytol 20: 355–364
Langridge J (1958) An osmotic mutant of *Arabidopsis thaliana.* Aust J Biol Sci 11: 457–470
Läuchli A (1976) Genotypic variation in transport. In: Lüttge U, Pitmam MG (eds) Encyclopedia of plant physiology, New Series, vol 2, part B. Springer, Berlin Heidelberg, New York, pp 372–393
Lin SC, Yuan LP (1980) Hybrid rice breeding in China. In: Innovative approaches to rice breeding. IRRI, Manila, Philippines, pp 35–51
Lyness AS (1936) Varietal differences in the phosphorus feeding capacity of plants. Plant Physiol 11: 665–688
Maas EV (1987) Salt tolerance of plants. In: Christie BR (ed) Handbook of plant science in agriculture, vol 2. CRC, Boca Raton, pp 57–75
Maas EV, Hoffman GJ (1976) Crop salt tolerance: Evaluation of existing data. In: Dregre HE (ed) Managing saline water for irrigation. Proc Int Conf Texas Tech Univ Lubbock, pp 184–198
Mahadevappa M, Ikehashi H, Noorsyamsi H, Coffman WR (1981) Improvement of native rices through induced mutation. IRRI Res Pap Ser 57, pp 7
Majumder ND, Borthakur DN, Rakshit SC (1989) Heterosis in rice under phosphorus stress, Indian J Genet 49: 231–235
Makmur A, Gerloff GC, Gabelman WH (1978) Physiology and inheritance of efficiency in potassium utilization in tomatoes grown under potassium stress. J Am Soc Hortic Sci 103: 545–549
Martinez CP (1976) Aluminum toxicity studies in rice (*Oryza sativa* L.) PhD Thesis Oregon State University, Corvallis, pp 113
McDaniel ME, Brown JC (1982) Differential iron chlorosis of oat cultivars. J Plant Nutr 5: 545–552
McDaniel ME, Dunphy DJ (1978) Differential iron chlorosis of oat cultivars. Crop Sci 18: 136–138
McDonald DJ, Gilmore EC, Stansel JW (1971) Heterosis for rate of gross photosynthesis in rice. Agron Abstr 63: 11–12
Mesdag J, Slootmaker LAJ (1969) Classifying wheat varieties for tolerance to high soil acidity. Euphytica 18: 36–42
Mesdag J, Slootmaker LAJ, Post J Jr (1970) Linkage between tolerance to high soil

acidity and genetically high protein content in the kernel of wheat *Triticum aestivum* L., its possible use in breeding. Euphytica 19: 163–174
Moeljopawiro S, Ikehashi H (1981) Inheritance of salt tolerance in rice. Euphytica 30: 291–300
Mugwira LM, Patel KL, Rao PV (1976) Lime requirement for triticale in relation to other small grains. Acta Agron Acad Sci Hung 25: 365–380
Murray JJ, Foy CD (1978) Differential tolerances of turfgrass cultivars to an acid soil high in exchangeable aluminum. Agron J 70: 769–774
Myers WM (1960) Genetic control of physiological processes. Consideration of differential ion uptake by plants. In: Caldecott RS, Snyder CA (eds) Radio isotopes in the biosphere. University of Minnesota, Minneapolis, pp 201–226
Naismith RW, Johnson MW, Thomas WI (1974) Genetic control of relative calcium phosphorus and manganese accumulation on chromosome 9 in maize. Crop Sci 14: 845–849
Narayana KK, Krishnaraj S, Sree Rangasamy SR (1990) Genetic analysis for salt tolerance in rice. In: Program Abstr 2nd Int Rice Genet Symp, 14–18 May 1990. IRRI, Manila, Philippines, pp 26
Nyborg M (1970) Sensitivity to manganese deficiency of different cultivars of wheat oats and barley. Can J Plant Sci 50: 198–200
Olsen SR (1972) Micronutrient interactions. In: Mortvedt JJ, Giordano PM, Lindsay WL (eds) Micronutrients in agriculture. Soil Science Society of America, Madison, WI, pp 243–264
Ouelette GJ, Genereux H (1965) Influence de l'inroxication manganique sur six variétiés de pomme de terre. Can J Soil Sci 45: 24–32
Pateña G, Vergara BS, BardhanRoy SK (1980) The use of rapid generation advance in rice breeding. Philipp J Crop Sci 5: 76 (Abstr)
Pearson GA, Ayers AD, Eberhard DL (1966) Relative salt tolerance of rice during germination and early seedling development. Soil Sci 102: 151–156
Polle E, Konzak CF, Kittrick JA (1978) Visual detection of aluminum tolerance levels in wheat by hemotoxylin staining of seedling roots. Crop Sci 18: 823–827
Pope DT, Munger HM (1953a) Heredity and nutrition in relation to magnesium deficiency chlorosis in celery. Proc Am Soc Hortic Sci 61: 472–480
Pope DT, Munger HM (1953b) The inheritance of susceptibility to boron deficiency in celery Proc Am Soc Hortic Sci 61: 481–486
Prestes AM, Konzak CF, Hendrix JW (1975) An improved seedling culture method for screenig wheat for tolerance to toxic levels of aluminum. Agron Abstr 67: 60
Prohaska KR, Fehr WR (1981) Recurrent selection for resistance to iron-deficiency chlorosis in soybeans. Crop Sci 21: 524–526
Purohit DC, Tripathi RC (1972) Performance of some salt-tolerant paddy varieties in Chambal commanded area of Rajasthan. Oryza 9: 19–20
Rajaram S, Matzenbacher R, Rosa S (1987) Developing bread wheats for acid soils through shuttle breeding In: CIMMYT research highlights 1986. CIMMYT, Mexico city, pp 37–47
Rana RS (1978) Wheat variability for tolerance to salt-affected soils. In: Gupta AK (ed) Genetics and Wheat improvement. Oxford and IBH, New Delhi, pp 180–184
Rana RS (1986) Breeding crop varieties for salt-affected soils. In: Chopra VL, Paroda RS (eds) Approaches for incorporating drought and salinity resistance in crop plants. Oxford and IBH, New Delhi, pp 25–55
Rana RS, Singh KN, Ahuja PS (1980) Chromosomal variation and plant tolerance to sodic and saline soils. In: Symp Pap Int Symp Salt Affected Soils, Central Soil Salinity Research Institute, Karnal, India pp 487–493
Ratanadilok N, Marcarian V, Schmalzel C (1978) Salt tolerance in grain sorghum. Agron Abstr 70: 160
Reddy N, Reddy K, Singh JM (1987) Breeding of a high yielding rice variety Deepak for problem soils. Fiji Agric J 49: 1–7
Reid DA (1970) Genetic control of reaction to aluminum in winter barley. In: Nilan RA

(ed) Barley genetics II. Proc 2nd Int Barley Genet Symp. Washington State University Press, Pullman, pp 409–413
Reid DA (1976) Genetic potentials for solving problems of soil mineral stress: aluminum and manganese toxicities in the cereal grains. In: Wright MJ (ed) Plant adaptation to mineral stress in problem soils. Cornell University Agricultural Experiment Station, Ithaca, NY, pp 55–64
Reide CR, Campos LAC (1988) Development of wheat cultivars with higher yield and adaptation to different agroclimatic conditions of Parana. In: Kohli MM, Rajaram S (eds) Wheat breeding for acid soils: review of Brazilian/CIMUYT Collaboration 1974–1986. CIMMYT, Mexico city, pp 26–38
Rhue D, Grogan CO, Stockmeyer EW, Everett HL (1978) Genetic control of Al tolerance in corn (*Zea mays* L.). Crop Sci 18: 1063–1067
Rosa O (1988) Current special breeding projects at the National Research Centre for Wheat (CNPT) EMBRAPA Brazil. In: MM Kohli, Rajaram S (eds) Wheat breeding for acid soils. Review of Brazilian/CIMMYT Collaboration 1976–1986. CIMMYT, Mexico city, pp 6–12
Rush DW, Epstein E (1981) Breeding and selection for salt tolerance by the incorporation of wild germplasm into a domestic tomato. J Am Soc Hortic Sci 106: 699–704
Sajjad MS (1990) Induction of salt tolerance in Basmati rice (Oryza sativa L.). Pertanika 13: 315–320
Salvi PV, Chavan KN (1983) Coastal saline soils of Maharashtra. J Indian Soc Coastal Agric Res 1: 21–26
Schettini TM, Gabelman WH, Gerloff GC (1987) Incorporation of phosphrous efficiency from exotic germplasm into agriculturally cropped germplasm of commonbean (*Phaseolus vulgaris* L.). In: Gabelman WH, Loughman BC (eds) Genetic aspects of plant mineral nutrition. Martinus, Dordrecht, pp 559–568
Senadhira D, Virmani SS (1987) Survival of some F_1 rice hybrids and their parents in saline soil. Int Rice Res Newsl 12(1): 14–15
Shafi M, Majid A, Ahmad M (1970) Some preliminary studies on salt tolerance of rice varieties. W Pak J Agric Res 8: 117–123
Shea PF, Gabelman WH, Gerloff GC (1967) The inheritance of efficiency in potassium utilization in strains of snapbeans (*Phaseolus vulgaris* L.). Proc AM Soc Hortic Sci 91: 286–293
Shukla UC, Raj H (1976) Zinc response in corn as influenced by genetic variability. Agron J 68: 20–22
Sinha TS, Bandyopadhyay AK (1984) Rice in coastal saline land of West Bengal India. In: Workshop on Research priorities in tidal swamp rice. IRRI, Manila, Philippines, pp 115–118
Slootmaker LAJ (1974) Tolerance to high soil acidity in wheat related species rye and triticale. Euphytica 23: 505–513
Spain JM (1976) Field studies on tolerance of plant species and cultivars to acid soil conditions in Colombia. In: Wright MJ (ed) Plant adaptation to mineral stress in problem soils. Cornell University Agricultural Experiment Station, Ithaca NY, pp 213–222
Spain JM, Francis CA, Howeler RH, Calvo F (1975) Differential species and varietal tolerance to soil acidity in tropical crops and pastures. In: Bornemisza E, Alvarado A (eds) Soil management in tropical America. Soil Sci Dep N Carolina State University, Raleigh, pp 308–329
Vergara BS, Pateña G, Peralta J, BardhanRoy SK, Eunus M (1980) Manual for rapid generation advance of rice. IRRI, Manila, Philippines
Virmani SS (1977) Varietal tolerance of rice to iron toxicity in Liberia. Int Rice Res Newsl 2(1): 4–5
Virmani SS (1979) Breeding rice for tolerance to iron toxicity. In: Proc 2nd Varietal Improvement Seminar WARDA S/76/7, 13–18 Sept 1976, Wet African Rice Development Association, Monrovia, Liberia, pp 156–173
Vose PB (1963) Varietal differences in plant nutrition. Herbage Abgstr 33: 1–13

Wall JR, Andrus CF (1962) The inheritance and physiology of boron response in the tomato. Am J Bot 49: 758–762

Wann EV, Hills WA (1973) The genetics of boron transport in tomato. J Hered 64: 370–371

Weiss MG (1943) Inheritance and physiology of efficiency in iron utilization in soybeans. Genetics 28: 253–268

West DW, Francois LE (1982) Effects of salinity on germination growth and yield of cowpea. Irrig Sci 3: 169–175

White MC, Decker AM, Chaney RL (1979) Differential cultivars tolerance in soybean to phytotoxic levels of soil zinc. I. Range of cultivar response. Agron J 71: 121–126

Yadav JSP (1979) Coastal saline soils of India. Bull 5 Central Soil Salinity Research Institute, Karnal. India

Yeo AR, Flowers TJ (1984) Mechanisms of salinity resistance in rice and their role as physiological criteria in plant breeding. In: Staples RC, Toenniessen GA (eds) Salinity tolerance in plants; strategies for crop improvement. Wiley, New York, pp 151–170

Yeo AR, Yeo ME, Flowers TJ (1988) Selection of lines with high and low sodium transport from within varieties of an inbreeding species; rice (*Oryza sativa* L.). New Phytol 110: 13–19

Yeo AR, Yeo ME, Flowers SA, Flowers TJ (1990) Screening of rice (*Oryza sativa* L.). genotypes for physiological characters contributing to salinity resistance and their relationship to overall performance. Theor Appl Genet 79: 377–384

Younis AF, Hatata MA (1971) Studies on the effects of certain salts on germination on growth of root and on metabolism. I. Effects of chlorides and sulphates of sodium potassium and magnesium on germination of wheat grains. Plant Soil 34: 183–200

Zaiter HZ, Coyne DP, Clark RB (1987) Genetic variation and inheritance of resistance of iron-deficiency chlorosis in dry beans. J Am Soc Hortic Sci 112: 1019–1022

Chapter 3

Physiological Criteria in Screening and Breeding

A. R. Yeo

3.1 Introduction

The great majority of plant improvements to date has been achieved on the basis of characteristics which can be clearly seen (e.g. shape and size) or which can be quantified directly (e.g. yield). So why might it be necessary to consider *physiological* characteristics which are not visible, and why is it necessary to measure anything other than yield, which is, after all, the prime objective? Breeding plants for problem soils, and indeed for tolerance to many environmental stresses, presents situations in which visible plant characteristics do not provide sufficient information for the plant breeder and in which yield is not an efficient index of the potential of parent lines. The purpose of this chapter is to examine the reasons for this.

Problem soils are those that suffer from one or more mineral excesses and/or deficiencies. Such soils often present a *complex* challenge to the plant, and the actual agent of damage cannot always be recognised. I widen my discussion to include water deficit, which commonly coexists with and exacerbates mineral stresses such as salinity. The tolerance of mineral stresses (and environmental stresses in general) is likely *also* to be complex because it may require the combination of a number of separate, and potentially independent, characteristics. This situation is in marked contrast to some other facets of plant breeding, for resistance to pests and diseases, for instance, where there is usually a single, recognisable causal agent. A "problem" soil usually means a soil area where a crop does not grow well, despite a need for it to be grown in that area. This arises basically because the range of edaphic conditions is very much larger than the ecological range of major crops in contemporary use. Often, therefore, there may not have been much opportunity for tolerance to particular conditions to have been developed in the cultivated species or in its immediate progenitors. Even if useful variation for the resistance of a soil problem *does* exist in the germplasm of a species (and it may not), the ecology of that species may not have provided selection pressure favouring the combination of such characters into a single genotype. In this event, plant breeding has a role if it can bring about the combination of the necessary traits into one variety, rather than treating the stress as a single character. This is a procedure for which the term pyramiding has been applied. Pyramiding entails some departure from normal

Monographs on Theoretical and Applied Genetics, Vol. 21
Ed. by A. R. Yeo and T. J. Flowers

plant-breeding practice for two important reasons. Firstly, simple screening is not an effective way of selecting parents. Secondly, characters without visual expression will need to be used in selecting parents and may be needed in handling breeding populations. This will increase the investment of resources in a breeding programme and so will affect the assessment of whether a breeding approach is appropriate to a particular problem.

3.2 Reasons for the Use of Physiological Selection

3.2.1 The Complexity of Tolerance

The underlying reason for the use of physiological characters in plant breeding is the variety of the physiological effects of, and plant responses to, environmental stresses. This is often complicated by the fact that more than one soil stress may occur simultaneously. As expanded in the examples to follow, there are few plant processes which are not affected; so it is not expected that tolerance has a single answer. A tolerant genotype can be expected to have more than a single adaptation.

Boursier and Läuchli (1989) found genotypic differences in chloride accumulation in leaf sheaths of sorghum, but this did not relate to observed differences in growth; they concluded that salt tolerance in any crop species is most likely correlated with the combination of several mechanisms acting together. Similarly, a crop growth simulation (for drought tolerance in potato) emphasised the complexity caused by the many plant processes involved, by the differences between instantaneous and cumulative effects, and by the strong genotype by environment interaction (Spitters and Schapendonk 1990). The correlation between salt content and survival was poor in rice, in spite of the fact that excessive salt in the leaves damages salt-sensitive crop species, because a number of other physiological, as well as morphological, characters are important to tolerance (Yeo et al. 1990).

Problem soils generally have a discontinuous spatial distribution; they occur in one part of a field but not another, for example. This contrasts with the temporal fluctuations which may occur in one place with some other inimical environmental variables (such as extremes of temperature and rainfall, and herbivory). Spatial variation offers less likelihood than temporal variation that the gene pools of crop species (which derive mostly from "favourable" conditions) have previously been exposed to any given unfavourable condition. This, in turn, reduces the expectation that genotypes optimised for problem soils already exist. Agricultural need and agricultural land degradation challenge crop species suddenly with environments not experienced (or experienced only unsuccessfully) during their evolutionary history. Whilst there is certainly no guarantee that tolerance to problems outside the ecological range of a species will be found within its germplasm, metal tolerance (see Chap. 7.1) does provide dramatic examples of latent ability to adapt to environments which are ecologically rare; although in some cases at least this may be due to the fortuitous different use of an existing gene product (see Chap. 6.2.4.2).

If, however, a *number* of genetically independent characteristics are required for tolerance and the characteristics required existed in the germplasm of a species, they could occur in a single genotype only as a result of sufficient natural selection pressure, or as a result of chance. For many of our crop species, many soil problems of agricultural importance are outside (or marginal to) their ecological range. Selection pressures may not have been very great. Consequently, there is no good reason to expect that a tolerant genotype is there for the finding. However, this is an essential prerequisite if the genotype needed for a breeding programme is to be identified through a simple screening procedure, the approach usually followed. The investment in a screening programme, when there are tens or hundreds of thousands of accessions of the major food species, is enormous, and may take years or tens of years to complete. A rational assessment of the probability of success needs therefore to be made.

If the variation needed does exist, but there has not been sufficient selection pressure to combine it, then the components of tolerance may be scattered throughout different genotypes. None of these is necessarily tolerant itself. This scenario is made more likely because so many of the genotypes now extant are cultivars developed for agronomic purposes at a time when the needs to use marginal land were less pressing. Selection for good performance in favourable conditions will have reduced variability within the germplasm and may have broken up genetic combinations that might have existed in ancestral races which may have been growing in different environments. A practical implication of this is that lack of success in initial screening does not mean that improvement cannot be achieved.

However, it would be hopeful indeed to expect that the handful of species that form the basis of our food supply (see Chap. 1.5) possesses the genetic variation, or even possessed it before domestication, to cope with the range of environments in which there is now pressure to grow crops.

3.2.2 Importing Tolerance from Wild Relatives

Most food production is concerned with a few tens of species and with but a few species of the Poaceae in particular. If these few species are to be adapted to a wide range of different environments, then new genetic information may be needed (Devine 1982). To facilitate this, there is currently interest in "wide crossing", whether this is brought about by conventional hybridisation (see Chap. 4.3 to 4.8) or by other means (see Chap. 4.8 and, e.g. Finch et al. 1990). The purpose of wide crossing is to introduce (additional) variation from a related species that may already possess tolerance to certain stresses as a result of its own ecological distribution. Wild relatives native to the target environment can often be found, but it is much less easy to make productive hybrids between them and cultivars of existing crops. It is not easy to obtain something agronomically useful from the cross, and there is no automatic expectation of being able to do so. To be successful, there is a need to recognise exactly what is useful to the crop genome in the alien genome, so that the rest can be deleted (or only selected parts merged in the first place). Otherwise the hybridisation is likely at best to dilute and at worst to corrupt the desirable qualities of the cultivated parent. The advantage of the genes for tolerance (if they have an *agricultural* advantage) may be outweighed by adverse effects of the foreign genetic material. Some pioneering work in this area was the hybridisation of the *ecologically* salt-tolerant *L. cheesmanii* with a cultivar of the domesticated tomato, *L. esculentum* (Rush and Epstein 1981). Although this cross was made many years ago, it has not resulted in commercial success. With repeated backcrossing, an increase in fruit yield at very high salinity was achieved, but at the cost of reduced yield compared with *L. esculentum* at lower salinities. The message is that salt tolerance per se (the *L. cheesmanii* accession grew on the shores of the Galapagos Islands) is not necessarily the best qualification for a donor parent. Unless specified genes can be identified, transferred and expressed in a crop background, or unless the complex of conditions that allow the sophisticated chromosome engineering used in species such as wheat (see Chap. 4.3) are met, then the most extreme parent may not be the first choice. For instance, other species of *Lycopersicon* possess a useful degree of salt tolerance, and are both closer to *L. esculentum* genetically and are more acceptable agronomically that *L. cheesmanii* (see Cuartero et al. 1992).

The mechanism by which a wild species survives a stressful environment may represent an extreme ecological adaptation; for instance, many are slow-growing perennials, which is quite incompatible with agricultural needs. The question is whether such species possess traits that can be transferred to crop species to improve their tolerance without compromising, unacceptably, their productivity. Since traits of interest may be concerned with such factors as nutrient discrimination and acquisition, and the distribution of nutrient and toxic ions within the plant, then the criteria for recognising these characters are likely to be physiological rather than visual. For instance, in wild species of *Cajanus* (pigeon-pea), salt tolerance was correlated with lower shoot sodium, and higher root sodium, concentrations than sensitive species; there were also increases in potassium concentration in tolerant species while this declined in sensitive species (Subbarao et al. 1990). A number of physiological traits associated with salt tolerance have been recognised in rice (Yeo et al. 1990). A sodium-potassium discrimination trait in wild species within the Triticeae is well documented (see Gorham et al. 1987) and breeding for salt tolerance within the Triticeae is discussed in detail in Chapter 4.

3.2.3 Measuring Stress Tolerance

There is an applied problem which is common to all breeding for stress tolerance. Tolerance usually has visible expression only in stressful conditions and so it is necessary to find a suitable field site or to devise a screening trial to assess both potential parents and breeding populations. Yield measurements are suited to the appraisal of a small number of selected lines, where performance is a statistical property of the population. It is not generally applicable to earlier stages of the breeding programme. Here, reliance has so far been based predominantly on the visible expression of growth (or growth reduction) or general appearance (visible damage; IRRI 1976), neither of which has any unique relationship with the stress to which tolerance is sought (Sect. 3.2.1). It is difficult to provide conditions (such as a droughted or saline field) in which to examine the worth of large numbers of genotypes. Aside from the uncertainties of climate, problem soils are notoriously heterogeneous (e.g. Malcolm 1983; Richards 1983). In the early history of a breeding population, where the plant material as well as the soil conditions may be highly variable, the plant breeder may only characterise *individuals* by their field performance with little confidence. There is even less chance of identifying genotypes which possess only components needed for tolerance; the impact one component has on overall performance is likely to be lost in environmental "noise". The alternative has been to conduct a greenhouse trial in which each genotype can be grown under more reproducible conditions with a series of "checks" to take account of seasonal and other variations in climate. The logistic problems of such an operation are that there are hundreds of thousands of accessions of the major cereal crops. Many of these accessions are highly heterogeneous, requiring extensive replication. Even in inbreeding species, those designated as cultivars may still retain genetic variability for salt tolerance (Flowers and Yeo 1981; Jones and Wilkins 1984), which may be due to residual heterozygosity (Yeo et al. 1988).

Characteristics which provide quantitative information, objectively over short time periods, can increase the precision and information content of measurements and decrease interference by environmental effects. Some physiological criteria may be screened rapidly and simply. Jones (1978) screened for tolerance of photosynthesis to saline and osmotic stress using leaf slices. Erdei et al. (1990) monitored polyamine accumulation in excised leaf segments using ionic and non-ionic osmotic stress to mimic salinity and water deficit, respectively; polyamine accumulation was seen in known drought- and salinity-tolerant wheat cultivars under non-ionic osmotic stress, but only in a salt-tolerant cultivar under ionic osmotic stress. Measurements of potassium efflux reflected membrane leakage and correlated with levels of enzymes that protect against oxidant damage, and also correlated with resistance to drought stress (Malan et al. 1990). Electrolyte leakage has also been used to monitor damage to membrane permeability caused by water stress and salinity to maize roots (Navari-Izzo et al. 1988). Shoot and root lengths and turgor pressure at the coleoptile stage correlated with data for flag leaf osmoregulation and for yield (Morgan 1988).

The other, and unique, contribution to be made by physiological selection criteria is that they can be specific enough to isolate individual traits from overall performance, which will be essential if tolerance has to be constructed as opposed to simply found.

3.2.4 Interaction Between Environmental Stresses

It is common that soil problems occur together or are further complicated by interaction with other environmental stresses. Lewis et al. (1989) observed an interaction between the nature of the source of nitrogen and the susceptibility of wheat to salinity. The susceptibility of maize to iron chlorosis depended upon the source of nitrogen; with nitrate as the source, a high pH in the apoplast led to iron precipitation and chlorosis, whilst with ammonium as the source an acid pH in the apoplast kept iron available (Mengel and Geurtzen 1988). Elevated carbon dioxide increased yield in sweet potato relative to yield in atmospheric carbon dioxide in both well-watered and water-stressed conditions (Bhattacharya et al. 1990) and interactions have been observed between, for instance, ozone and drought (Bender et al. 1991). Acid saline and sodic soils are clear cases of the simultaneous occurrence of two quite different problems; salinity combined with low and high pH respectively – both of which may be directly damaging in their own right and as a consequence of the effect of extreme pH upon nutrient availability. There are commonly differences in genotypic performance in saline soils and in soils where salinity is combined with acidity or alkalinity (e.g. Bandyopadhyay 1988; Marassi et al. 1989). Many saline soils lie in the arid and semi-arid tropics where interaction between salinity and water deficit is inevitable. Interaction between salinity and any environmental factor that affects the rate of transpiration (light, temperature, wind, vapour pressure deficit, carbon dioxide depletion and so on) are well documented (see, e.g. Flowers et al. 1977, 1986).

In all, such interactions exacerbate the general situation outlined in Section 3.2.1. In the absence of information which allows the causes and effects of different stresses to be separated, the only course of action would be to seek varieties that yield adequately over a range of adverse environments and soil conditions. However, the limitations and restrictions which would be imposed by making yield stability the primary criterion, form a potent argument for increasing understanding of the interaction between plants and environmental stress.

3.2.5 Limitations to the Use of Yield as a Selection Criterion

Cultivars of many crops which are resistant to soil problems have been produced by the conventional approach of empirical selection and testing of yield performance over a range of years and environments using appropriate statistical methods (see Chap. 2). The level of investment and the time taken are considerable and are dictated by the statistical property of the selection criterion (yield) for which there are a number of reasons (Blum 1989):

- The inheritance of yield is complex, involving physiological, biochemical and morphological characteristics whose genetics are largely unknown and whose exact association with plant productivity is often unclear.
- The factors causing yield reduction in different environments are difficult to apportion and are generally amalgamated into the concept of stability of yield (Blum 1989). The component that reflects resistance to one particular aspect of the environmental regime (such as water or salinity) is difficult to isolate.
- Stress conditions that reduce yield also increase the environmental component of variation. This reduces the heritability of yield in these conditions and consequently the efficiency of the selection process.

For this combination of reasons, Blum (1989) regards selection for drought resistance based on yield as inefficient, too costly, and giving diminishing returns as further improvement is sought; he concludes that physiological selection criteria must supplement yield evaluation. A similar argument should apply to other stresses which act in similarly general ways and even more strongly to stresses acting in concert, such as when salinity and aridity are combined.

There is also an increasingly prevalent view that the development of a small number of cultivars possessing yield stability across a range of environments is not the way forward. This is because there are so many local variations in problem soils, which in turn interact with local climate, that a wide range of cultivars targeted at the local conditions is preferable. In these circumstances, yield in a particular environment will gain greater importance than yield stability over a range of environments. Another very important consideration is that if a variety is intended *only* for growth in stressful conditions, then its performance in ideal conditions becomes less important. Some of the priority given usually to those agronomic characters associated with maximal yield potential could, if necessary, be relaxed, an option which is not available in the development of widely planted cultivars.

The next section illustrates the role which physiological studies can play in helping define the properties needed in a tolerant cultivar, and in providing the methodology to employ physiological criteria in selection and breeding in parallel with conventional techniques. Salinity and water deficit are taken as the main examples.

3.3 Salinity

3.3.1 Basic Problems

At the lower salinities, uptake of salt (used here to refer to those ions dominating the saline soil) occurs because there is a mass flow of saline solution towards the root as a result of the transpiration stream, and the root is imperfect at discriminating between saline and nutrient ions. Entry of salt may occur by leakage across the membranes of the root (Kuiper 1985), by poor selectivity of carriers or ion channels at some point along the radial pathway to the xylem (Läuchli 1984; Gorham et al. 1987) or by leakage along an apoplastic pathway from the outside to the xylem (Yeo et al. 1987). Selectivity between nutrient and non-nutrient ions is usually considerable, even in salt-sensitive species, but it may be insufficient to cope with a saline environment. The more or less efficiently the plant uses water, the lesser or greater the flow of saline solution towards the plant for each unit of growth.

Salt damage arises through a combination of reasons. Even at low salinities, the external concentration of salt is much greater than that of nutrient ions, so the concentration of salt reaching the xylem may be considerable despite a high value for the selectivity (for instance, between sodium and potassium). The transpiration stream is a one-way transport towards the actively transpiring leaves, which are also the most important part of the plant in terms of net photosynthesis. The increasing salt concentration in the expanded leaves leads to their premature death (Munns and Termaat 1986) because the salt concentration in the symplast becomes inhibitory, and/or because the concentration of salt in the apoplast causes water deficit in the leaf (Oertli 1968). In a salt-sensitive crop (rice) and at lower salinities at least, excessive accumulation of salt in the expanded leaves is the initial site of damage (Yeo et al. 1991) and the primary cause of loss of agricultural productivity. The impact made by the salt is amplified by the localised concentration effects in the small apoplastic volume (Flowers et al. 1991) which accounts for the drought symptom of leaf-rolling observed at low *external* salinities.

3.3.2 Salt Exclusion

In conditions of relatively low salinity the major adaptations expressed by plants, and consequently the major objectives of plant breeding, are concerned with minimising the uptake of salt and limiting the damage the salt which does enter causes within the plant. There is generally an inverse correlation, between and within salt-sensitive species, between salt content of the leaves and salt tolerance of the plant. Exclusion of salt may be a sufficient objective for low salinities and the most sensitive crops where salt damage is almost entirely the consequence of internal accumulation, but it is important to note that exclusion may prejudice tolerance of higher salinities. Since the overall objective of salt exclusion is to limit the concentration in the leaves, then a rapid growth rate (dilution by growth) will achieve a result similar to restriction of uptake.

The question is whether or not the product of transpiration and xylem concentration causes the salt concentration in a leaf to rise fast enough to cause its decreased productivity or premature senescence. Provided that the salt load can be accommodated without a reduction in the photosynthetic output of the leaf, then that external salinity will not have an adverse effect. If photosynthesis is reduced, then productivity will be reduced. Net photosynthesis is affected by leaf expansion rate, leaf area and leaf duration, as well as by photosynthesis and respiration per unit leaf area. Since any reduction in shoot growth reduces the volume in which to accommodate newly arriving salt, a growth effect once begun is liable to become progressively worse (Munns and Termaat 1986), leading to a catastrophic failure when root selectivity is affected by the inadequate supplies of carbohydrate from the shoot (Yeo and Flowers 1986).

Properties related only to minimising salt uptake may be assisted by mechanisms that can moderate some of the consequences of excessive uptake. These are concerned with the distribution of the salt within the plant. For instance, putting all the salt in the old leaves causes less damage, on aggregate, than having similar concentrations everywhere (Yeo and Flowers 1982). In both salt-sensitive (e.g. *Phaseolus* and *Zea*) and salt-tolerant (e.g. *Puccinellia*) species, sodium may be scavenged from the xylem to provide some limited protection to the photosynthesising leaves (Jacoby 1965; Besford 1978; Stelzer 1981).

These are all *avoidance* mechanisms, concerned with keeping low levels of salinity in at least the laminae of leaves that should be photosynthetically active. An important factor is that the mechanisms are *visually* indistinguishable because the only observable feature is whether or not the leaves suffer visible damage. It cannot be deduced from appearance alone whether plants growing without damage in a saline soil possess genetic information for one or more tolerance mechanisms, or are simply the most vigorous.

Avoidance mechanisms can include adaptations which are suited to growth in a more saline environment. Even plants which are naturally salt-tolerant sometimes possess one more sophisticated avoidance mechanism: salt glands.

3.3.3 Conditions Requiring Osmotic Adjustment

As concern moves to salinities which are high enough to require osmotic adjustment to the external water potential, then the rules are changed. The plant must accumulate or synthesise *something* which is osmotically active or it will wilt. The quantities needed may not be as great as expected, because there may be no advantage in maintaining the turgor pressure observed before salinisation (Munns 1988). Avoidance alone will not, however, lead to a productive plant (though it may contribute to an ecologically tolerant one). Mechanisms and resources are necessary for (1) utilising salt within the cells for osmotic adjustment or (2) for obtaining sufficient non-saline ions for osmotic adjustment in the presence of salt or (3) for producing organic solutes. Utilising salt for osmotic adjustment at average concentrations in the cell that are greater than could be tolerated in the cytoplasmic compartment, as observed in many halophytes, is the only mechanism which *exploits* the saline environment, by separating the osmotic and metabolic roles of ions and so utilising salt to maintain water potentials and growth (Flowers et al. 1977, 1986). Few species of current commercial importance exhibit these characteristics, which is understandable because they adapt plants to highly saline environments at the expense of productivity in non-saline environments.

3.3.4 Characteristics Needed in Salt-Tolerant Plants

The characteristics needed to achieve salinity tolerance in plants can thus be categorised according to the salinity at which improvement is sought. They are introduced in sequence of increasing external salinity. It is a fact, with far-reaching consequences, that the mechanisms conferring resistance of low salinity and tolerance of high salinity can be mutually exclusive.

3.3.4.1 Control of Salt Uptake

The uptake of salt occurs through those processes normally involved in the accumulation of nutrients. Such uptake might be minimised by:

- a membrane lipid composition that minimises passive permeability;
- minimising the "bypass-flow" in regions of the root with an undifferentiated or damaged endodermis and/or exodermis;
- maximising the selectivity of the carriers and ion channels that are responsible, respectively, for the active and passive transport of ions across the membrane of the root and into the xylem;
- maximising water use efficiency because, other things being equal, this minimises the potential salt load per unit of new growth and thus the likely concentration in the tissue;
- maintaining vigorous growth and/or the continual replacement of lost leaves; because dilution-by-growth is one way of minimising salt damage and because leaf replacement is essential to offset losses.

3.3.4.2 Limiting the Damage That Excessive Ion Uptake Causes

This may be achieved through the ability of the plant to:

- localise the saline ions in old leaves and so protect those leaves at the peak of their photosynthetic productivity;
- compartmentalise ions within the leaf to minimise the dehydration effects of apoplastic accumulation: this means that accumulating ions into leaf cell is less damaging than leaving them in the cell walls;
- remove excess salt from the leaves (using structures ranging from simple unicellular microhairs to complex multicellular glands).

3.3.4.3 Osmotic Adjustment

Osmotic adjustment is necessary for growth at high salinities. This can be achieved in a variety of ways:

- possession of a large capacity for nutrient acquisition combined with high selectivity between for instance sodium and potassium and between chloride and nitrate;
- synthesis of organic solutes, such as sugars and organic acids, to adjust both cytoplasm and vacuole;
- compartmentalisation of saline ions in the vacuoles to minimise concentrations in the symplasm;
- synthesis of a "compatible cytosolute", an organic solute that has a neutral (or even protective) effect upon metabolism even at high (molar) concentrations, to balance the low water potential that will occur in the vacuole;
- possession of salt glands and/or the capacity to develop succulence to allow for long-term regulation of leaf salt concentration. Since salt will continue to arrive in the transpiration stream, the salt concentration in the leaf would rise continually unless the salt can be excreted or diluted by succulence. The former has much greater capacity but the latter provides some margin of safety where the balance between growth and ion uptake is good.

There are, then, three categories of physiological characteristic that are important in the tolerance of salinity by plants. What is apparent is that there is no single answer to tolerating salt, even if the case is simplified to coastal saline soils where the influence is maritime (and so predominantly sodium chloride) and independent of complication by extremes of pH.

3.3.5 Agricultural Versus Ecological Advantage

Mechanisms of salt tolerance can also be separated according to whether they may be of agricultural advantage or only of advantage in terms of ecological survival. The ionic conditions within the cytoplasm have to be maintained within certain limits because of both general (lyotropic) and specific (activation) effects upon proteins. The activation requirements of protein synthesis generally require about 100 mM potassium combined with a Na:K ratio of not much more than unity (see Leigh and Wyn Jones 1984). There are some indications of slightly wider tolerances in halophytes (Flowers and Dalmond 1992) but overall there is a ceiling of perhaps 200–300 mM on the concentration of inorganic ions irrespective of their ionic species. If a plant is unable to compartmentalise salts in the vacuole (that is, to have a substantially higher concentration and different relative composition of salts in the vacuole than in the cytoplasm), then the requirements for the cytoplasm must be met by the cell as a whole. This will constrain the extent to which sodium can be substituted for potassium, and impose a limit on the degree of osmotic adjustment that can be achieved with inorganic solutes. In an ecological sense this is irrelevant, provided that the advantage of survival in saline conditions is sufficient in its own right and there is little need to grow fast to compete with other individuals. If a low growth rate is acceptable, then the plant can rely upon acquiring sufficient potassium from the environment and/or diverting a large proportion of its photosynthate into osmotic adjustment of the vacuolar volume. However, this will not satisfy the agricultural need for a substantial growth rate. Compartmentation allows halophytes to use sodium chloride to fill the large vacuoles and to conserve potassium and organic solutes for the much smaller cytoplasmic volume. These plants are capable of substantial growth rates in highly saline conditions. Thus, from an agricultural perspective, the mechanism of tolerance is as important as tolerance determined by the ability to survive.

3.3.6 Examples Where Knowledge of Physiological Mechanisms Would Aid Selection for Salt Tolerance

The consequence of the foregoing is that it is very difficult to devise a screening procedure that will provide the right answer: a genotype suited to agricultural productivity at a given soil salinity. Some examples of the difficulties associated with selection based on appearance or overall performance, about which a knowledge of the physiological mechanisms involved would allow a more reasoned choice, are given in the following paragraph.

3.3.6.1 Where Tolerance Is Accidental

For example non-dwarf rices (tall, traditional landraces) are generally more salt-resistant than dwarfed varieties (modern cultivars). This is attributable at least in part to dilution of salt by vegetative vigour rather than to specific genetic information for salinity tolerance, hence their salt tolerance is accidental. The non-dwarf plant type is not favoured agriculturally because of poor yield potential and liability to lodging. If a non-dwarf plant is used as a 'donor' of salt tolerance in a cross with a dwarf, improved variety then there is a likelihood that its advantage will simply be lost. The tolerance of new material produced from a "traditional" non-dwarf parent is at best intermediate between the parents (see Chap. 2.4.3.1). This has slowed progress in breeding for salinity tolerance in rice, even though the germplasm collection has been systematically screened (Akbar 1986 and Chap. 2.2.1).

3.3.6.2 Where Tolerance Is Not a Single Character

Tolerance may require a number of characters and no particular genotype possesses them all. There is therefore no ideal genotype to locate. Parents with separate traits cannot reliably be located on their overall performance because single traits do not have sufficient impact on the overall performance to make that genotype outstanding. A screening trial based on overall performance will conclude that there is insufficient variation within the species and a breeding approach to the problem will be abandoned, or other pathways such as mutation breeding and interspecific (or intergeneric) hybridisation will be advocated. It may still be simpler and more economic to remain with conventional breeding and genotypes with agricultural acceptability, but use physiological information to supplement the selection tools.

3.3.6.3 Where Tolerance Is Incidental

A character may be advantageous for an incidental reason. For instance, water use efficiency should increase salt tolerance, but it may best be sought in a drought-tolerant parent. Without knowing that the trait might be useful, it would not be included in a breeding programme for salt tolerance.

3.3.6.4 Where a Mechanism Is As Important As Tolerance Itself

In attempting to increase the level of salinity that can be tolerated, it would seem logical to start with a genotype that is already tolerant of low salinity, and aim to improve upon it. Knowledge of plant physiology suggests that this may, in fact, be the worst possible starting place because the mechanism of tolerance of low salinity may preclude that genotype from tolerance to a higher salinity.

3.3.6.5 Where Screening/Selection Pressure Is Severe

If a screening trial is too severe, it will be biased towards identifying survivors, which may not be agriculturally useful, and miss genotypes that combine acceptable combinations of tolerance and growth rate.

In each case, more information is needed than the sum total of the plants' performance.

3.4 Drought

Although drought is strictly an environmental stress rather than a soil problem, it is discussed here because parallels are so often drawn between physiological responses to drought and salinity (e.g. Turner and Passioura 1986), because some mechanisms involved in drought tolerance also contribute to tolerance of soil problems, and because water deficit is commonly an additional factor in problem soils.

Drought resistance is a complex characteristic with effects at different stages of plant development, and whose effect varies with intensity and duration (Poehlman 1987). In the same way that salinity tolerance mechanisms need to be chosen to target the environment in which an improvement is sought, so is the case for drought. The severity of drought determines whether the only aim is to maintain photosynthesis under stress, or whether tolerance of desiccation and consequences of reduced transpiration, such as overheating, must be included. The duration and frequency or probability of drought affects decisions about the extent to which the mechanism by which water stress is tolerated can be at the expense of yield potential.

There is a well-established dependency between economic yield, as a component of biomass production, and the volume of water transpired by the crop during the growing period (Passioura 1983). In rice, for example, this is generally 250–300 g of water transpired per gram of carbon gained (Yoshida 1981). Crops using the C_4 pathway of photosynthesis are considerably more efficient in their use of water, and exhibit very high growth rates. Plants capable of Crassulacean acid metabolism (CAM) are very efficient in their use of water, but the achievement of this through stomatal closure during daylight marks CAM as a survival strategy and growth rates are not generally compatible with agricultural objectives. Any mechanism which simply *prevents* water loss will be at the expense of yield. To maintain yield potential it is necessary to maintain transpiration under stress (Blum 1989); yield can be maintained with reduced transpiration only if the efficiency of water use can be increased. External factors also enter into the argument, and transpiration under stress can be a *disadvantage* if it will deplete the water that will be available to the crop before the vegetative growth can be converted into the harvested component.

The physiological basis of tolerance of water deficit can be divided into a number of categories involving different mechanisms, more than one of which is likely to be needed in a crop variety that is productive under water-limited conditions.

3.4.1 Efficient Use of Water

There are characters of plant growth and habit which affect water loss but which are not concerned with the physiology of plant water relations. For example, early canopy closure restricts evaporative losses from the soil surface, and rapid attainment of a leaf area index in excess of two is the best prospect of conserving this water (Passioura 1986). Early maturation is another factor which may allow the crop to mature, or at least to get past the especially critical stages of anthesis, while the water supplies in the soil are still adequate. Anatomical characters such as glaucousness in wheat and a thick, waxy cuticle in sorghum contribute to a reduction in the heat load on the leaf and to resistance of desiccation.

Water-use efficiency is perhaps the most important physiological character contributing to the tolerance of drought. Useful variation is found within crop species but has been difficult to determine routinely. Instantaneous gas exchange (carbon dioxide and water vapour) measurements and experiments with plants in pots reflect only some (different) components of crop water use. The ability to determine water-use efficiency a posteriori by stable isotope analysis (Farquhar and Richards 1984) has opened up the possibility of using this as a selection criterion. The basis of the method is discrimination during photosynthesis between ^{12}C and ^{13}C. High water-use efficiency is associated with low partial pressures of carbon dioxide in the leaf, and this means a lower discrimination against ^{13}C by the carboxylation reactions. Condon et al. (1987) showed that the ranking for carbon isotope discrimination of 27 genotypes of wheat at two contrasting sites was similar, suggesting that genotype by environment interaction was small for this character. Consistent with this view, a number of studies have shown that carbon isotope discrimination was a heritable character (see Farquhar et al. 1989). The advantages of such measurements are twofold. Firstly they give an integrated picture of plant water use efficiency *within the canopy*, something that instantaneous gas exchange measurements cannot do. Secondly, all that must be done in the field is to sample some leaves from the plant for analysis elsewhere. Although the procedure is essentially "high-tech", it does not require that these measurements are made in remote sites. There is a problem, though. The natural abundance of ^{13}C is small and the discrimination during carbon fixation is also small, so the carbon isotope ratio measurements need to be made with very high resolution by a slow procedure on an expensive machine. The cost and throughput-time of such mass spectrometry would limit use at present to choosing parents and evaluating elite lines rather than evaluating breeding populations.

A further factor affecting efficiency of water use is respiratory loss. On a whole-plant, 24-h basis, respiratory losses are 30–50% of photosynthetic carbon uptake, even in favourable conditions (McCree 1986), something which is liable to be underestimated in studies of single leaves (McCree 1986). Wilson (1984) has selected within ryegrass for low rates of respiration, which may be a beneficial character if it reflects increased efficiency in the use of respiratory energy, and low respiration rates do not prejudice the plant's performance in favourable conditions.

3.4.2 Exploitation of Soil Moisture

Obtaining water relies upon the ability of root growth to exploit fully the water resources in the soil. Obtaining water requires not only deep roots but efficient extraction. Any carbon allocated to developing and maintaining an increased investment in roots will, however, offset some of the yield increase gained by accessing extra water. Axial conductance may also be important. A monocotyledonous plant without secondary thickening will need more root axes to carry the water. Differences in root density at different soil depths in wet and dry soils have been observed (e.g. IRRI 1982). The observed distribution of roots could arise from the combination of a vigorous rooting system combined with differential death of surface roots in a dry soil. Alternatively, there could be a positive response to the perception of soil water deficit (perception by plants of soil water deficit is discussed below). This would seem an important distinction to resolve.

Evaluation of rooting pattern, involving as it does the excavation of mature plants in the field, is clearly not amenable to use in breeding populations. It could, however, be used to evaluate, as potential parents, a moderate number of genotypes suggested by their field reputation. There is a need for more rapid and convenient selection procedures for rooting characteristics. This will not be straightforward because "convenient" systems (hydroponics and pots) do not mimic at all well the problems of growing through a drying soil. Potential strategies are to characterise the properties of roots that would enable them to penetrate drying soils and also to investigate signals that might be used to alter the pattern of root development. There has been recent interest in the question of root-to-shoot signalling and its role in reducing stomatal conductance and leaf growth rate when the *root* experiences water deficit (e.g. Termaat et al. 1985; Munns 1988; Sharp and Davies 1989). If a signalling system *has* evolved to give advance warning to the shoot of water regime problems being experienced by the root, then root growth could conceivably utilise the same signalling system itself.

3.4.3 Leaf Water Relations

In a droughted or saline soil the water potential is decreased and water availability may become less than that required to replace transpirational loss from the plant. In the absence of any form of adjustment in the leaf to compensate for this, the turgor pressure in the leaf will fall. How much this matters has been questioned by a number of recent studies which suggest that turgor pressure does not directly control cell expansion or stomatal conductance (see Munns 1988). However, it is essential for productivity, and in many cases for survival, that the leaf does not go on losing water until it rolls or wilts. In this state the leaf will be suffering net carbon loss and, although the enforced stomatal closure will tend towards recovery of turgor, the leaf will have periods when is vulnerable to heat and desiccation damage. Events *in the leaf* that may maintain turgor are, in the short term, osmotic adjustment or reduced stomatal conductance and, in the longer term, a reduction in cell volume.

Osmotic adjustment, as strictly defined, means a net increase in the quantity of osmotically active solutes within the cell. The increase in solutes lowers the solute potential which assists the maintenance of a turgor pressure. The value of osmotic adjustment has often been regarded as self-evident, but Munns (1988) challenges this on two grounds. Firstly that osmotic adjustment (to drought at least) must utilise photosynthetic products that could otherwise be used for growth. Secondly, that osmotic adjustment to maintain the turgor pressure existing before stress is wasteful, because maintenance of that turgor is not necessary. Observed osmotic adjustment often agrees well with the resources made available by growth reduction both in saline (see Yeo 1983) and drought (see Munns 1988) conditions. However, this may also be brought about by a change in the pattern of assimilate partitioning between organs (Blum et al. 1988). Furthermore, a *positive* correlation between osmotic adjustment and yield is also observed empirically (Morgan 1983; Blum 1989). There are two reasons why this may be. Firstly, a reduction in turgor may be acceptable, but a loss of turgor is not; if conditions are such that a *loss* of turgor would otherwise occur, then anything which prevents this is likely to be of advantage. Secondly, growth is not the only consequence of maintaining turgor and keeping the stomata open, and carbon balance is not the only role of leaf gas exchange. Another essential role of transpiration is in the control of leaf temperature and in some environments there is an inverse correlation between canopy temperature and yield under stress. The transpiring leaves cool as a result of the latent heat of vaporisation of water and this can be detected readily by thermal imaging techniques (Blum 1989). High leaf temperatures indicate stomatal closure and may be associated with lethal overheating of the leaves, a common observation under drought and a potential complication of the interpretation of leaf damage under salinity. The consequences are that even if osmotic adjustment consumes photosynthetic carbon, there are conditions in which it is likely to result in a survival advantage, and it could still result in a yield advantage in stressed conditions.

A decrease in cell volume can lead to an increase in solute concentration without a net increase in the quantity of osmotic solutes. Small cells also maintain turgor at lower values of water potential than do larger cells (Cutler et al. 1975; Castro-Jimenez et al. 1989). Reduction in cell size may be associated with a reduced leaf expansion rate and a reduced leaf area, both of which will reduce the potential for photosynthetic carbon fixation. All mechanisms of restoring leaf turgor are liable to conflict with carbon fixation and so with yield. They do differ greatly in the time scales over which they act and respond. Changes in stomatal conductance are rapid and reversible, osmotic adjustment is slower (and the organic solutes could later be used for growth), changes in cell and leaf size are more final.

3.4.4 Selection for Drought Tolerance

As for salinity, there is no single description of what is a drought stress and no single answer to the characters required of a crop genotype to cope with water deficit. Intensity, duration and probability of water deficit, as well as other climatic factors (temperature and radiation in particular) all affect the benefit, and the penalty, conferred by any mechanism of drought tolerance.

The overall appearance of the plant does not convey enough information to deduce which mechanism or mechanisms are being expressed in a particular genotype. In the simplest case, genotypes that have a field reputation may perform poorly in pot experiments because the latter mainly test the ability to reduce water loss and tolerate desiccation and cannot easily indicate differences in rooting behaviour which may have been the main feature in the field. The mechanism is material because different approaches to the avoidance or resistance of water deficit are appropriate to different patterns of water deficit in the field. A combination of characteristics is likely to be necessary; roots which extract water efficiently, some mechanism for keeping leaf turgor at least above zero, and in many cases the tolerance of desiccation and the avoidance or tolerance of high leaf temperatures. All of these may need to be recognised and handled separately.

One major difference between breeding for drought and breeding for salinity is that water deficit of some description is an integral part of the environment of most plants. Even in well-watered soils leaves experience water deficit at high irradiance and high transpirational demand. Consequently, the expectation of discovering in crop species variation, and useful combinations of characteristics, is very much greater in the case of drought than it is in the case of salinity.

3.5 Physiological Selection Procedures

It is first necessary that any screening procedure correctly duplicates or represents the soil problem in the field. This may require extensive characterisation of the site to determine whether the edaphic stress is, for instance, an acid pH per se, Al toxicity, Mn toxicity, a deficiency arising from the soil pH, or some combination of these things (Devine 1982). On a practical basis, characterisation may also determine whether or not a breeding programme is the most economic strategy. Some programme of fertilisation or soil amelioration could prove less costly than a breeding programme.

A second requirement, that is much more difficult to meet, is to maximise the genetic versus environmental effects seen in the screening trial. The problems here of working in the field are enormous. To take salinity as an example, the difficulties fall into two categories. The first is the heterogeneity of the environment itself. Saline soils are very patchy, often a mosaic of quite abrupt changes on a fine scale. The second is interaction with environmental factors. Anything which affects the transpiration rate affects salt tolerance (Sect. 3.2.4).

These difficulties can be approached in two ways. Firstly by moving the trial from the field into at least partially controlled conditions. This means having a detailed knowledge of the nature and level of the stress being targeted, because it must be reproduced in glasshouse or other artificial situations. The second is to increase the information content of whatever assay is performed. Salinity damage in rice is usually scored in the field, and in the glasshouse, by visually assessing the proportion of the leaves of the plant that are damaged (IRRI 1976). However, leaves are liable to be damaged for different reasons in different genotypes at different salinities; for instance, by salt accumulation, by external water deficit, by heat damage if either of the former have caused stomatal closure in a leaf that does not exhibit rolling, and even by nutritional disorders attributable to the salinity of the growth medium. Any of these may interact with other environmental variables. It follows that the appearance of the plant does not tell us as much about salt transport as would a measurement of leaf sodium content or Na : K ratio. A "physiological" measurement provides more, and less equivocal, information.

3.5.1 Advantages of Using Physiological Criteria

The advantages of applying physiological criteria to plant selection are three-fold. Firstly, as a complement to or a replacement for visual assessment, such criteria can supply more objective and quantitative information than visual assessment itself. Secondly, it makes possible the handling of component traits of complex characters, which may have no visibly recognisable expression. Thirdly, by providing objective criteria, evaluations can be made by a wider range of personnel than those assessments that rely on the skill and experience of the breeder which is only accumulated over many years.

3.5.2 Constraints

Visual assessment has only labour costs, is capable of being applied to large numbers of plants quickly, and places no reliance on technology. Physiological measurements take longer, are more expensive, and need equipment. Crop improvement programmes commonly take place in remote locations with very limited resources.

3.5.3 Prospects

To use physiological criteria in a breeding programme, the breeding programme has to be viewed in its three separate stages: (1) the selection of parents, (2) the handling of breeding populations and (3) the evaluation of elite breeding material. Sophisticated techniques can be applied to the selection of parents because timing is not a critical constraint. Routine screening of the germplasm banks of major crops, even by observation alone, may take decades. It is not necessary that the search for parents be carried out in the field or even in less developed countries. This is a situation where National and International Centres and advanced institutions in developed countries have a clear comparative advantage. Once parents have been selected, the resultant breeding populations have to be handled in a way which is appropriate to the breeding system and breeding method. This will, in most cases, mean that large numbers of individual plants must be grown, and assessed, over a period of time at the breeding site. For inbreeding species this generally means six or more generations. For many soil problems this will take place in the national breeding programmes of countries in Africa, in Asia and in Latin America. The laboratory procedures which are appropriate ways to choose parents are clearly not appropriate to the handling of breeding populations and will not be useful again until a manageable number of elite populations has been selected. Since the present trend is that much of this material will be a pre-breeding stock for use in local breeding programmes, rather than a released cultivar, then there is a strong case for knowing what characters it may donate.

There are three possible approaches as regards whether physiological selection continues to have a role with breeding populations and, if so, what this role would be.

The first is to rely subsequently upon the breeders' judgement. The role of physiological selection will have been to suggest parents or combinations of parents *which would not otherwise have been chosen.* The rationale will have been that these parents (or combinations of parents) provide a better prospect of developing the required tolerance than parents chosen on other criteria of merit (visible characters or field reputation) and so provide better starting material than would otherwise have been used. It will then be an act of faith that the characters of the original parents survive the criteria of conventional selection. There are few data available, but from the experiences of my own laboratory I would have to counsel caution.

The second approach is to aim to translate the original selection procedures into something which is sufficiently cheap, simple, rapid and reliable to use in the field to evaluate a breeding population of perhaps thousands of plants. As examples, ion-selective electrodes and portable porometers and infra-red gas analysers are equipment which can meet some or all of these criteria. Such techniques could allow assessment of a range of characters relating to salinity and to mineral nutrition and to plant productivity as affected by them. Leaf and canopy temperature measurement can give information about stomatal conductance and vulnerability to heat damage: this can be achieved by remote sensing (Smith et al. 1989). This is, therefore, a practicable avenue at the present time for certain techniques. Limited laboratory facilities at breeding stations could expand the range considerably if analyses can be made on samples from field plants. For instance, the techniques using leakage (see above and Cakmak and Marschner 1988; Vasque-Tello et al. 1990) require quite simple equipment. Zhang and Davies (1990) found that xylem ABA was a very sensitive indicator of soil drying and the development of immunoassays has made the quantification of ABA a technically simple and extremely rapid procedure.

The third avenue revolves on the concept of genetic markers; something which is visible, or which can be visualised, and is linked to the "invisible" character whose presence in sought. Genetic markers began with linkage maps of morphological characters, were extended with the use of isozymes, and have now been greatly extended with the development of restriction fragment length polymorphism (RFLP) maps of some crop species. At the present time, RFLP technology is not cheap, nor that rapid, and not particularly simple; but as an area in which technological development can be anticipated, and as a suitable candidate for technology transfer, it has much in its favour. The great advantage is that a single technique could be applied to a whole range of breeding problems, *once a trait and RFLP markers for it had been identified.* This contrasts with the wide range of laboratory techniques that would be needed to measure directly the diverse characteristics associated with salinity, drought, nutrient deficiencies and toxicities, and so on.

References

Akbar M (1986) Breeding for salinity tolerance in rice. In: IRRI (ed) Salt-affected soils of Pakistan, India and Thailand. International Rice Research Institute, Manila, Philippines, pp 39–63

Bandyopadhyay AK (1988) Performance of some acid tolerant rice varieties in two acid saline soils of Sunderbans. Int Rice Res Newsl 13(3): 19

Bender J, Tingey DT, Jager HJ, Rodecap KD, Clark CS (1991) Physiological and biochemical responses of bush bean (*Phaseolus vulgaris*) to ozone and drought stress. J Plant Physiol 137: 565–570

Besford RT (1978) Effect of replacing nutrient potassium by sodium on uptake and distribution of sodium in tomato plants. Plant Soil 58: 427–432

Bhattacharya NC, Hileman DR, Ghosh PP, Musser RL, Bhattacharya S, Biswas PK (1990) Interaction of enriched CO_2 and water stress on the physiology of and biomass production in sweet potato grown in open-top chambers. Plant Cell Environ 13: 933–940

Blum A (1989) Breeding methods for drought resistance. In: Jones HG, Flowers TJ, Jones MB (eds) Plants under stress. Cambridge University Press, Cambridge, pp 197–216

Blum A, Mayer J, Golan G (1988) The effect of grain number per ear (sink size) on source activity and its water-relations in wheat. J Exp Bot 39: 106–114

Blum A, Ramaiah S, Kanemasu ET, Paulsen GM (1990) The physiology of heterosis in *Sorghum* with respect to environmental stress. Ann Bot 65: 149–158

Boursier P, Läuchli A (1989) Mechanisms of chloride partitioning in the leaves of salt-stressed *Sorghum bicolor* L. Physiol Plant 77: 537–544

Cakmak I, Marschner H (1988) Increase in membrane permeability and exudation in roots of zinc deficient plants. J Plant Physiol 132: 356–361

Castro-Jimenez Y, Newton RJ, Price HJ, Halliwell RS (1989) Drought stress responses of *Microseris* species differing in nuclear DNA content. Am J Bot 76: 789–795

Condon AG, Richards RA, Farquhar GD (1987) Carbon isotope discrimination is positively correlated with grain yield and dry matter production in field-grown wheat. Crop Sci 27: 996–1001

Cuartero J, Yeo AR, Flowers TJ (1992) Selection of donors for salt tolerance in tomato using physiological traits. New Phytol 121: 63–69

Cutler JM, Rains DM, Loomis RS (1975) The importance of cell size in the water relations of plants. Plant Physiol 40: 255–260

Devine TE (1982) Genetic fitting of crops to problem soils. In: Christiansen MN, Lewis CF (eds) Breeding plants for less favourable environments. Wiley, New York, pp 143–174

Erdei L, Trivedi S, Takeda K, Matsumoto H (1990) Effects of osmotic and salt stresses on the accumulation of polyamines in leaf segments from wheat varieties differing in salt and drought tolerance. J Plant Physiol 137: 165–168

Farquhar GD, Richards RA (1984) Isotopic composition of plant carbon correlates with water-use efficiency of wheat genotypes. Aust J Plant Physiol 11: 539–552

Farquhar GD, Wong SC, Evans JR, Hubick KT (1989) Photosynthesis and gas exchange. In: Jones HG, Flowers TJ, Jones MB (eds) Plants under stress. Cambridge University Press, Cambridge, pp 47–70

Finch RP, Slamer IH, Cocking EC (1990) Production of heterokaryones by the fusion of mesophyll protoplasts of *Porteresia coarctata* and cell suspension-derived protoplasts of *Oryza sativa*: a new approach to somatic hybridisation in rice. J Plant Physiol 136: 592–598
Flowers TJ, Dalmond D (1992) Protein synthesis in halophytes: the influence of potassium, sodium and magnesium in vitro. Plant Soil 146: 153–161
Flowers TJ, Yeo AR (1981) Variability in the resistance of sodium chloride salinity within rice varieties. New Phytol 88: 363–373
Flowers TJ, Troke PF, Yeo AR (1977) The mechanism of salt tolerance in halophytes. Annu Rev Plant Physiol 28: 89–121
Flowers TJ, Hagibagheri MA, Clipson NCW (1986) Halophytes. Q Rev Biol 61: 313–337
Flowers TJ, Hajibagheri MA, Leach RP, Rogers WJ, Yeo AR (1989) Salt tolerance in the halophyte *Suaeda maritima*. In: Tazawa M, Katsumi M, Masuda Y, Okamoto H (eds) Plant water relations and growth under stress. Yamada Science Foundation, Osaka, pp 173–180
Flowers TJ, Hajibagheri MA, Yeo AR (1991) Ion accumulation in the cell walls of rice plants growing under saline conditions: evidence for the Oertli hypothesis. Plant Cell Environ 14: 319–325
Gorham J, Hardy CA, Wyn Jones RG, Joppa LR, Law CN (1987) Chromosomal location of a K/Na discrimination character in the D genome of wheat. Theor Appl Genet 74: 584–588
IRRI (1976) Standard evaluation system for rice. International Rice Research Institute, Manila, Philippines
IRRI (1982) Drought resistance in crop plants with emphasis on rice. International Rice Research Institute, Manila, Philippines
Jacoby B (1965) Sodium retention in excised bean stems. Physiol Plant 18: 730–739
Jones HG (1978) Screening for tolerance of photosynthesis to osmotic and saline stress using rice leaf-slices. Photosynthetica 13: 9–14
Jones MP, Wilkins DA (1984) Screening for salinity tolerance by rapid genertion advance. Int Rice Res Newsl 9: 9–10
Kuiper PJC (1985) Environmental changes and lipid metabolism of higher plants. Physiol Plant 64: 118–122
Läuchli A (1984) Salt exclusion: an adaptation of legumes for crops and pastures under saline conditions. In: Staples RC, Toenniessen GH (eds) Salinity tolerance in plants; strategies for crop improvement. Wiley, New York, pp 171–187
Leigh RA, Wyn Jones RG (1984) A hypothesis relating critical potassium concentrations for growth to the distribution and functions of this ion in the plant cell. New Phytol 97: 1–13
Lewis OAM, Leidi EO, Lips SH (1989) Effect of nitrogen source on growth response to salinity stress in maize and wheat. New Phytol 111: 155–160
Malan C, Greyling MM, Gressel J (1990) Correlation between CuZn superoxide and glutathione reductase, and environmental and xenobiotic stress tolerance in maize inbreds. Plant Sci 69: 157–166
Malcolm CV (1983) Wheatbelt salinity, a review of the salt land problem of South-Western Australia. Western Australian Department of Agriculture, Perth
Marassi JE, Collado M, Benavidez R, Arturi MJ, Marassi JJN (1989) Performance of selected rice genotypes in alkaline, saline, and normal soils and their interaction with climate factors. Int Rice Res Newsl 14(6): 10
McCree KJ (1986) Whole-plant carbon balance during osmotic adjustment to drought and salinity stress. Aust J Plant Physiol 13: 33–43
Mengel K, Geurtzen G (1988) Relationship between iron chlorosis and alkalinity in *Zea mays*. Physiol Plant 72: 460–465
Morgan JM (1983) Osmoregulation as a selection criterion for drought tolerance in wheat. Aust J Agric Res 34: 607–614
Morgan JM (1988) The use of coleoptile response to water stress to differentiate wheat genotypes for osmoregulation, growth and yield. Ann Bot 62: 193–198
Munns R (1988) Why measure osmotic adjustment? Aust J Plant Physiol 15: 717–726

Munns R, Termaat A (1986) Whole-plant responses to salinity. Aust J Plant Physiol 13: 143–160

Navari-Izzo F, Izzo R, Bottazzi F, Ranieri A (1988) Effects of water stress and salinity on sterols in *Zea mays* roots. Phytochemistry 27: 3109–3115

Oertli JJ (1968) Extracellular salt accumulation, a possible mechanism of salt injury in plants. Agrochimica 12: 461–469

Passioura JB (1983) Roots and drought resistance. Agric Water Manage 7: 265–280

Passioura JB (1986) Resistance to drought and salinity: avenues for improvement. Aust J Plant Physiol 13: 191–201

Poehlman JM (1987) Breeding field crops. Van Nostrand Reinhold, New York

Richards RA (1983) Should selection for yield in saline regions be made on saline or non-saline soils? Euphytica 32: 431–438

Rush DW, Epstein E (1981) Breeding and selection for salt tolerance by the incorporation of wild germplasm into a domestic tomato. J Am Soc Hortic Sci 106: 699–704

Sharp RE, Davies WJ (1989) Regulation of growth and development in plants growing with a restricted supply of water. In: Jones HG, Flowers TJ, Jones MB (eds) Plants under stress. Cambridge University Press, Cambridge, pp 71–94

Smith RCG, Prathapar SA, Barrs HD (1989) Use of a thermal scanner image of a water stressed crop to study soil variability. Remote Sens Environ 29: 111–120

Spitters CJT, Schapendonk AHCM (1990) Evaluation of breeding strategies for drought tolerance in potato by means of crop growth simulation. Plant Soil 123: 193–203

Stelzer R (1981) Ion localisation in the leaves of *Puccinellia peisonis*. Z Pflanzenphysiol 103: 27–36

Subbarao GV, Johansen C, Jana MK, Kumar Rao JVDK (1990) Physiological basis of differences in salinity tolerance of pigeonpea and its related wild species. J Plant Physiol 137: 64–71

Termaat A, Passioura JB, Munns R (1985) Shoot turgor does not limit shoot growth of NaCl-affected wheat and barley. Plant Physiol 77: 869–872

Turner NC, Passioura JB (eds) (1986) Plant growth, drought and salinity. CSIRO, Melbourne

Vasque-Tello Y, Zuily-Fodil Y, Thi ATP, De Silva JBV (1990) Electrolyte and Pi leakages and soluble sugar content as physiological tests for screening resistance to water stress in *Phaseolus* and *Vigna* species. J Exp Bot 41: 827–838

Wilson D (1984) Identifying and exploiting genetic variation in the physiological components of production. Ann Appl Biol 104: 527–536

Yeo AR (1983) Salinity resistance: physiologies and prices. Physiol Plant 58: 214–222

Yeo AR, Flowers TJ (1982) Accumulation and localisation of sodium ions within the shoots of rice varieties differing in salinity resistance. Physiol Plant 56: 343–348

Yeo AR, Flowers TJ (1986) Salinity resistance in rice (*Oryza sativa* L.) and a pyramiding approach to breeding varieties for saline soils. Aust J Plant Physiol 13: 161–173

Yeo AR, Yeo ME, Flowers TJ (1987) The contribution of an apoplastic pathway to sodium uptake by rice roots in saline conditions. J Exp Bot 38: 1141–1153

Yeo AR, Yeo ME, Flowers TJ (1988) Selection of lines with high and low sodium transport from within varieties of an inbreeding species; rice (*Oryza sativa* L.). New Phytol 110: 13–19

Yeo AR, Yeo ME, Flowers SA, Flowers TJ (1990) Screening of rice (*Oryza sativa* L.) genotypes for physiological characters contributing to salinity resistance, and their relationship to overall performance. Theor Appl Genet 79: 377–384

Yeo AR, Lee K-S, Izard P, Boursier PJ, Flowers TJ (1991) Short- and long-term effects of salinity on leaf growth in rice (*Oryza sativa* L.). J Exp Bot 42: 881–889

Yoshida S (1981) Fundamentals of rice crop science. International Rice Research Institute, Manila, Philippines

Zhang J, Davies WJ (1990) Changes in the concentration of ABA in xylem sap as a function of changing soil water status can account for changes in leaf conductance and growth. Plant Cell Environ 13: 227–285

Chapter 4
Cytogenetic Manipulations in the Triticeae

B. P. Forster

4.1 Introduction

As the name implies, cytogenetics evolved from the merger of interests in cytology and genetics. A historical account of the evolution of cytogenetics is given by Schulz-Schaeffer (1980). The essence of the science is the chromosome and how chromosome behaviour determines gene transmission and recombination. Conventional cytogenetics is based on the study of whole or large segments of chromosomes which can be detected using a light microscope by various staining procedures. In recent years, cytogeneticists have added biochemical and molecular techniques to the science as chromosomes are studied in ever greater detail.

The tribe Triticeae (Hordeae) of the grass family Poaceae (Gramineae) contains some of the most important small grain cereal crops. This includes bread wheat, macaroni wheat, barley, rye and triticale, the synthetic wheat/rye hybrid. This chapter concentrates on cytogenetic research aimed at improving the salt tolerance of bread wheat; this is for several reasons. Firstly, salinity is a serious problem which restricts bread wheat production in many regions of the world. Secondly, bread wheat is an ideal crop for cytogenetic manipulations as it has relatively large chromosomes which can be studied individually by the use of specially constructed aneuploid stocks. Thirdly, chromosome recombination between the genomes of bread wheat and other members of the Triticeae can be induced allowing the transfer of genes into bread wheat from related species. Although some of the chromosome manipulations described are specific to bread wheat, many of the techniques can be exploited in cytogenetic studies of other crops. It should be noted that the genetic make-up of bread wheat can be exploited to study the genetic control of stress tolerance in related species, including the crops rye and barley, and some discussion of this is given.

4.2 Background to Wheat and Salinity

Saline soils cover a large part of the Earth's surface. It has been estimated that some 950 million hectares of the world land surface – about the area of Canada –

Monographs on Theoretical and Applied Genetics, Vol. 21
Ed. by A. R. Yeo and T. J. Flowers

are now affected by salt. The problem is most acute in arid and semi-arid regions where low rainfall, high surface evaporation, weathering of native rocks and clearing of perennial vegetation combine to produce saline soils. Irrigation and poor drainage can exacerbate the situation by producing secondary salinity. For example, in Pakistan 2 tonnes of salt are added to every irrigated hectare annually. The United Nation's Food and Agriculture Organisation estimated that half of the world's irrigated farms have been damaged by salinity (Pearce 1987). Large-scale irrigation and drainage schemes and chemical treatments of the soil are non-biological methods of managing saline soils. However, the scale of the problem often renders these solutions too costly; Pakistan has spent $140 million on reclaiming saline and waterlogged land (Stoner 1988). Alternative strategies are therefore required.

One biological approach is to introduce crops which are naturally adapted to saline soils (Charnock 1988). Many of these species are trees and shrubs (e.g. *Prosopis juliflora* and *Phoenix dactylifera*) of low economic value and which take many years to reach maturity and produce a commercial crop. Shrubs such as *Suaeda fructicosa* or various *Atriplex* species grow well in saline soils, but can be used only for fuel wood or fodder. There has, therefore, been increasing pressure, economical, political and social to improve the salt tolerance of staple crop species.

Crop tolerance to salinity is a major concern of the International Wheat and Maize Improvement Centre (CIMMYT, Mexico), which aims to produce cultivars for developing countries. In 1984, links were established between CIMMYT, the Centre for Arid Zones Studies (CAZS, UK) and the Plant Breeding Institute (PBI, UK). The aim was to capitalise on biochemical/physiological (CAZS) and genetic (PBI) expertise in accelerating CIMMYT's breeding effort for greater salt tolerance in wheat, particularly in the development of cultivars for the Indian subcontinent.

Some variation for tolerance to salt has been demonstrated in world collections of bread wheat, *Triticum aestivum* (Quershi et al. 1980; Kingsbury and Epstein 1984; Sayed 1985), and potential exists for small improvements by conventional breeding methods (Rana 1986). Genes which influence tolerance to salt have been located in the D genome of bread wheat (Gorham et al. 1987; Shah et al. 1987). These genes originated in *Aegilops squarrosa*, the donor of the D genome to the hexaploid bread wheat; this species and other D genome-carrying species therefore widen the accessible pool for salt tolerance genes. Since these genes can be found in a genome common with wheat, recombination and gene flow to the crop from the wild species is possible by the normal processes of meiosis; this strategy is being pursued by Farooq et al. (1989) and Munns et al. (1990). However, more distant relatives of wheat offer much more potent genes for tolerance to salt. These species belong to the *juncea* and *elongata* complexes of the genus *Thinopyrum* (*Agropyron*). Both *Th. bessarabicum* (*Ag. junceum*) and *Th. elongatum* (*Ag. elongatum*) have been used in salt tolerance research for wheat improvement.

4.3 Cytogenetics of Wheat

In order to understand the various cytogenetic techniques available for transferring genes into wheat from alien sources it is first important to understand the genetic make-up of wheat which allows such chromosome engineering.

Bread wheat, *Triticum aestivum*, is a hexaploid species having 21 chromosome pairs (2n = 6x = 42). These chromosomes are grouped into three genomes designated A, B and D. The three genomes originate from three closely related diploid species which hybridised to form hexaploid wheat (see Miller 1987). Because the A, B and D genomes are closely related, genes on one genome are frequently found to have homoeoalleles on the same linkage groups in the other two genomes. A consequence of this triplication of genes is that hexaploid wheat is genetically buffered, a feature fundamental to the genetic manipulations used in alien gene transfer. One way in which this genetic buffering is exploited is in the development of aneuploid genetic stocks such as in a monosomic series. A complete monosomic series in a wheat (2n = 41, rather than 42) consists of 21 lines, each of which lacks one chromosome of each of the 21 chromosome pairs. The deficiency is buffered by compensating genes on homoeologous chromosomes. Monosomic series are of great value in both pure and applied genetic studies; a catalogue of wheat monosomic series is given by Worland (1988). Monosomic lines can be used to locate genes to specific chromosomes and to transfer individual chromosomes from one wheat line into another (Law et al. 1987). A monosomic series in a modern CIMMYT wheat cultivar, Glennson 81, was developed at PBI as an integral part of the cytogenetics programme to improve the salt tolerance of wheat.

Another important feature of wheat cytogenetics is the capacity of wheat to tolerate additional chromosomes and entire genomes. A great range of wheat/alien hybrids and amphiploids has been produced in which an alien genome has been added to wheat (Maan and Gordon 1988). On crossing amphiploids back to wheat and subsequent repeated backcrossing and selfing, it is possible to isolate each chromosome of the alien species in a wheat genetic background to produce wheat/alien disomic chromosome addition lines (Gale and Miller 1987; Shepherd and Islam 1988). Disomic chromosome addition lines (2n = 44, Fig. 4.1) are of great genetic interest, as they can be used to locate genes specific to the isolated chromosome (Gale and Miller 1987). Ditelosomic chromosome addition lines which contain an added telosomic pair, either the long or the short arm (2n = 42 + 2t), are invaluable in locating genes directly to specific chromosome arms. Chromosome addition lines can be used in conjunction with monosomic lines to replace chromosomes of wheat with related alien chromosomes (Gale and Miller 1987). Such substitution lines have a stable karyotype (2n = 42, Fig. 4.1) and can be developed into commercial cultivars: Zorba and Weique wheats, which carry a substitution of chromosome 1R of rye for 1B of wheat, were developed in such a way and released in eastern Europe (Zeller 1973).

Chromosome translocation lines can also be produced in wheat (Fig. 4.1). These often occur spontaneously in progeny of plants which undergo unbalanced meiotic pairing, although the mechanism of production is unclear. It is thought that univalents (unpaired chromosomes) at meiosis have a tendency to break at their centromeres and that the separated chromosome arms can reunite; if more than one univalent does this, hybrid Robertsonian-type translocated chromosomes may result (Sears 1973; Fig. 4.1). Translocation lines can be of great commercial importance, the prime example being the translocation of the short arm of chromosome 1R of rye onto the long arm of 1B of wheat

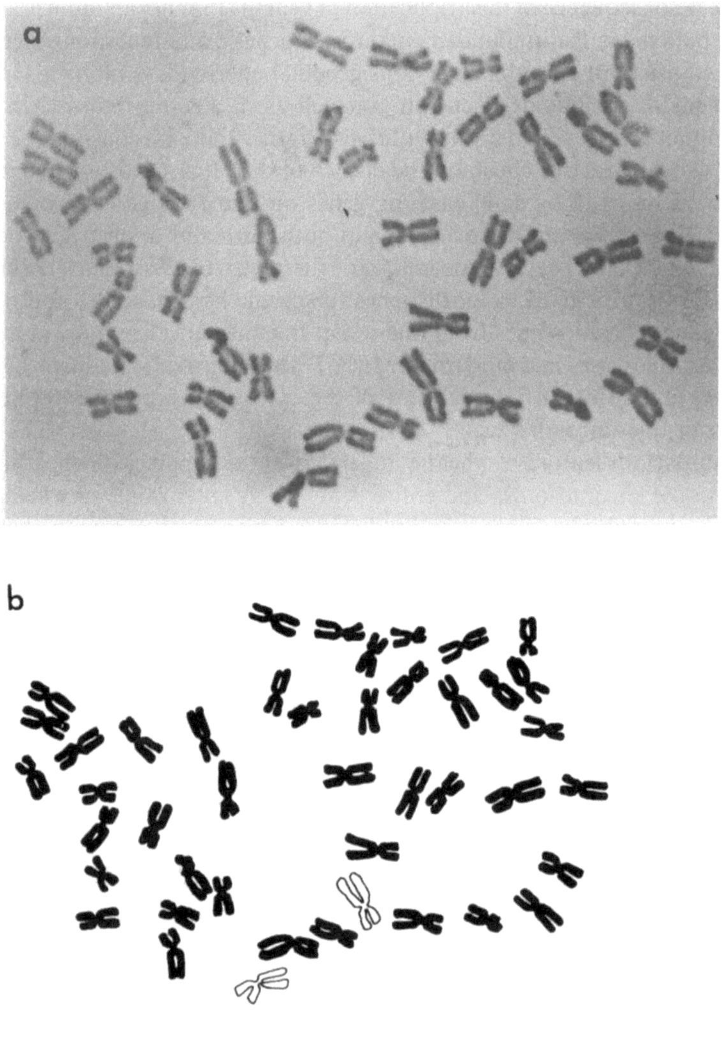

Fig. 4.1a,b

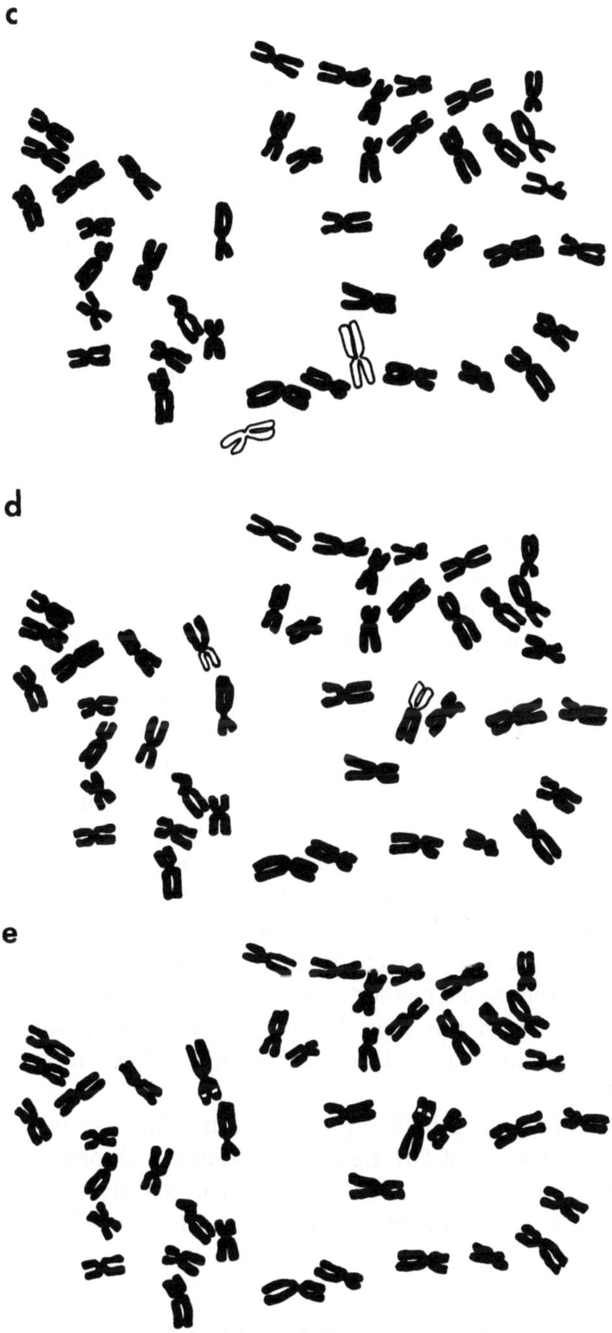

Fig. 4.1a–e. Examples of chromosome engineering in wheat. **a** 44 chromosomes of a wheat/alien disomic chromosome addition line. **b** Diagram of addition line showing the alien chromosome pair (2n = 44). **c** Substitution line (2n = 42). **d** Translocation line (2n = 42). **e** Recombinant line

(Zeller 1973; Mettin et al. 1973). These 1BL/1RS lines carry resistance genes to several diseases (stem rust, *Sr31*; leaf rust, *Lr26*; stripe rust, *Yr9*, and powdery mildew, *Pm8*: McIntosh 1988), and have increased yield potential (Rajaram et al. 1983). Consequently, they are present in many cultivars throughout the world. A disadvantage of both substitution and translocation lines is that a large amount of genetic material is introduced and this can carry deleterious as well as desirable genes. For instance, although 1RS carries genes for disease resistance, it also confers poor bread-making quality. Similarly, the wheat/*Thinopyrum elongatum* translocation in the wheat cultivar Agatha which confers leaf rust resistance has the deleterious effect of producing yellow flour (Knott 1968).

The next step in transferring alien genes into wheat is the production of recombinant lines. Here alien chromosomes present in the addition or substitution lines are recombined with those of wheat so that only a small portion of the alien DNA (the piece carrying the desired trait) is introduced (Fig. 4.1). The effect of this sophisticated transfer is to produce a near normal wheat, which benefits from an acquired trait, but which lacks any deleterious character present in the donor species. The production of wheat/alien recombinant lines can be achieved by manipulating yet another feature of wheat genetics, the *Ph1* gene. This is not difficult, but requires crossing with specific genetic stocks and progeny testing.

One, at first, surprising feature of wheat cytology is the fact that at meiosis wheat behaves as a diploid species and forms 21 bivalents. From the above description of wheat evolution it might be expected, because of the close relationship between the A, B and D genomes, that meiosis in hexaploid wheat would comprise 7 hexavalents and not 21 bivalents. The strict bivalent pairing observed is due to the expression of the *Ph1* gene, located on the long arm of chromosome 5B, which actively suppresses associations between homoeologous chromosomes so that only homologous chromosomes pair (Okamoto 1957). The *Ph1* gene poses a major obstacle to the recombination between wheat and alien chromosomes. There are two genetic ways around this. The first is simply to use wheat lines which are deficient for chromosome 5B and hence deficient for *Ph1*. The second method is to use genetic stocks which carry either a deletion or mutation for this gene. In both systems, association between homoeologous chromosomes are induced and recombination can be affected (Gale and Miller 1987).

The genetic buffering of wheat, its ability to tolerate aneuploidy and the manipulation of *Ph1* are central features to chromosome engineering of wheat. These features have been exploited in a systematic approach to identifying, locating and transferring alien genes into wheat for salt tolerance.

4.4 Transfer into Wheat of Alien Genes for Tolerance to Salt

Thinopyrum bessarabicum (sand couch) was identified by CAZS as a potential donor species for salt tolerance genes (Wyn Jones et al. 1984). *Thinopyrum bessarabicum* is a diploid species native to sand dunes of the Black Sea and is tolerant to prolonged exposure to high salt concentrations. Early physiological studies showed that it is able to limit the accumulation of sodium (Na) and chloride (Cl) in its leaves (Gorham et al. 1985, 1986a). The regulation of salt transport from roots was also thought to be an important mechanism limiting salt accumulation in mature leaves of non-halophytic crop plants including wheat (Joshi et al. 1979; Greenway and Munns 1980). Other factors in favour of using *Th. bessarabicum* were its ploidy ($2n = 2x = 14$, E^bE^b) and its crossability with wheat (Alonso and Kimber 1980), although no natural recombination with wheat occurs. Being diploid, nuclear genes are confined to just seven chromosome pairs. If a higher polyploid species had been chosen, it would have been necessary to screen 14 or 21 chromosomes for genes controlling tolerance to salt.

The cytogenetic programme at PBI began with a cross between *Th. bessarabicum* and wheat *Triticum aestivum* cv. Chinese Spring monosomic for chromosome 5B (Fig. 4.2). The mono-5B line was emasculated and pollinated 2–4 days later with pollen from the wild species. Developing hybrid embryos were excised 14 days after pollination and cultured on orchid agar to ensure survival; hybrid embryos of wide crosses frequently die before reaching maturity in vivo. Two hybrid karyotypes were produced from the cross. One carried the euploid complement of 21 wheat and 7 *Th. bessarabicum* chromosomes ($n = 28$), the other was found to be deficient, nullisomic for chromosome 5B ($n = 27$). Meiotic studies of the hybrids showed that pairing occurred between wheat and alien chromosomes in the 5B-deficient, *Ph1*-deficient, hybrid, but not in the euploid hybrid (Forster and Miller 1985). This was a significant finding, as it indicated that gene transfer from *Th. bessarabicum* to wheat could be achieved through meiotic recombination in the absence of *Ph1* and that the novel variation could be fixed on restoration of *Ph1*.

Both hybrid genotypes proved to be sterile. The nulli-5B hybrids also failed to respond to chromosome-doubling attempts using colchicine, and eventually died. Hence it was not possible to introgress genes through backcrossing procedures in the absence of *Ph1*. The euploid, 28 chromosome hybrid did respond to colchicine-doubling and produced a fertile wheat/*Th. bessarabicum* amphiploid ($2n = 8x = 56$, $AABBDDE^bE^b$), and this formed the foundation of subsequent work (Fig. 4.2).

Hydroculture tests showed that the wheat/*Th. bessarabicum* amphiploid is substantially more tolerant to salt stress than the wheat parent, Chinese Spring, and a range of reputedly salt-tolerant wheat cultivars (Gorham et al. 1986b; Forster et al. 1987). The amphiploid was able to grow and complete its life cycle at 250 mol m^{-3} NaCl, whereas the threshold level of wheat is between 175–200 mol m^{-3} NaCl (Fig. 4.3). Laboratory results were confirmed by field trials in salt-affected lands in Saskatchewan, Canada. It was concluded that *Th. bessarabicum* genes for salt tolerance are dominant and expressed in a wheat genetic background.

The next step involved locating salt tolerance genes to specific chromosomes in the E^b genome of *Th. bessarabicum*. Wheat/*Th. bessarabicum* disomic

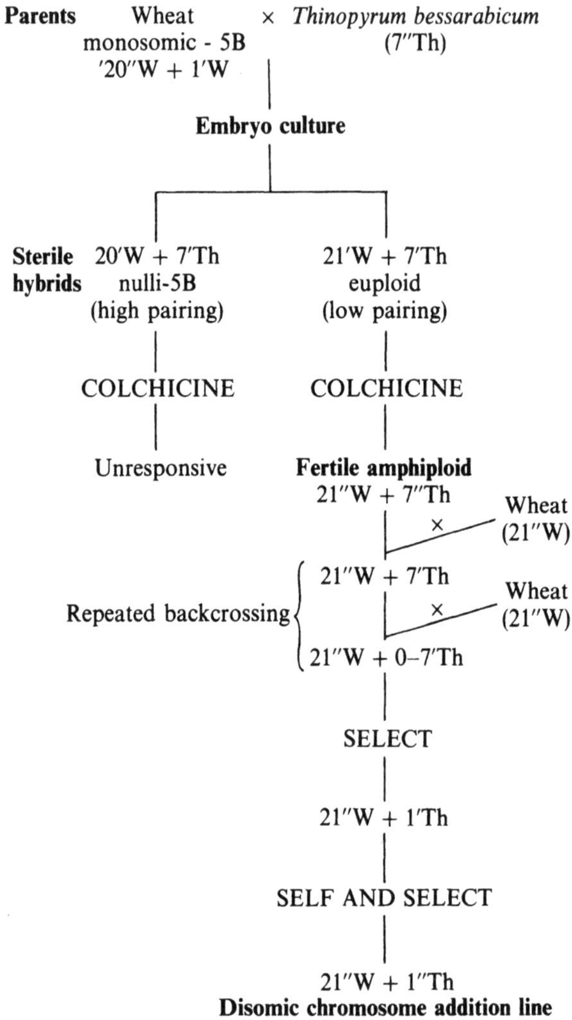

Fig. 4.2. Crossing procedure used to produce wheat/*Thinopyrum bessarabicum* disomic chromosome addition lines

chromosome addition lines in Chinese Spring were produced at PBI (B. P. Forster and T. E. Miller; Fig. 4.2) and at the University of Missouri, USA (G. Kimber and C. Alonso). The combined collection contained 17 distinct lines, many of which carried intra-genomic translocations between *Thinopyrum* chromosomes. The salt tolerance of the 17 lines was tested. The majority showed the same level of tolerance as the wheat parent, Chinese Spring, but two lines gave different responses. One line carrying a pair of $2E^b$ chromosomes was found to be more susceptible to salt damage than wheat, another addition line carrying the $5E^b$ chromosome showed greater

Fig. 4.3. Salt tolerance of wheat (CS), *Thinopyrum bessarabicum* and the wheat/*Thinopyrum bessarabicum* amphiploid. The plants are growing in a nutrient solution containing 250 mol m^{-3} NaCl, *from left to right*; three plants each of wheat, the amphiploid and *Thinopyrum bessarabicum*. Wheat cannot tolerate this concentration of salt and is dead, the amphiploid is able to complete its life cycle and is maturing, it produces plump, viable seed, *Thinopyrum bessarabicum* is slower-growing and is at anthesis. (Courtesy of the Cambridge Laboratory, John Innes Centre, Norwich, UK)

tolerance. These two lines were then retested along with Chinese Spring, the amphiploid, and also wheat lines tetrasomic for homoeologous group 2 and group 5 chromosomes. This was done in order to assess and compare the effects of the added genes from *Th. bessarabicum* with those of wheat at similar dose levels (Forster et al. 1988). The order of tolerance ranked: amphiploid > Chinese Spring > $2E^b$ addition > tetra-2A = tetra-2D > tetra 2B at 150 mol m^{-3} NaCl and amphiploid > $5E^b$ addition > Chinese Spring = tetra-5A > tetra 5D = tetra 5B at 200 mol m^{-3} NaCl (Fig. 4.4). It was concluded that although genes influencing tolerance to salt are probably distributed throughout the E^b genome, chromosomes $2E^b$ and $5E^b$ are of importance and that $5E^b$ carries a gene(s) with major effects.

Attempts are now underway to transfer the $5E^b$ character into wheat. The $5E^b$ addition line has been crossed onto wheat lines monosomic for the chromosomes 5A, 5B and 5D to develop chromosome substitution lines. An added bonus in the case of $5E^b$ (5B) substitutions is the induction of recombination between $5E^b$ and 5A, and $5E^b$ and 5D. Chromosome translocation lines can also be generated by the backcross procedure. It is important at this stage that targeted genes are transferred into a suitable wheat genotype. The need for a monosomic series in a modern wheat cultivar adapted to Indian subcontinental conditions was anticipated, and a monosomic series had been produced in the

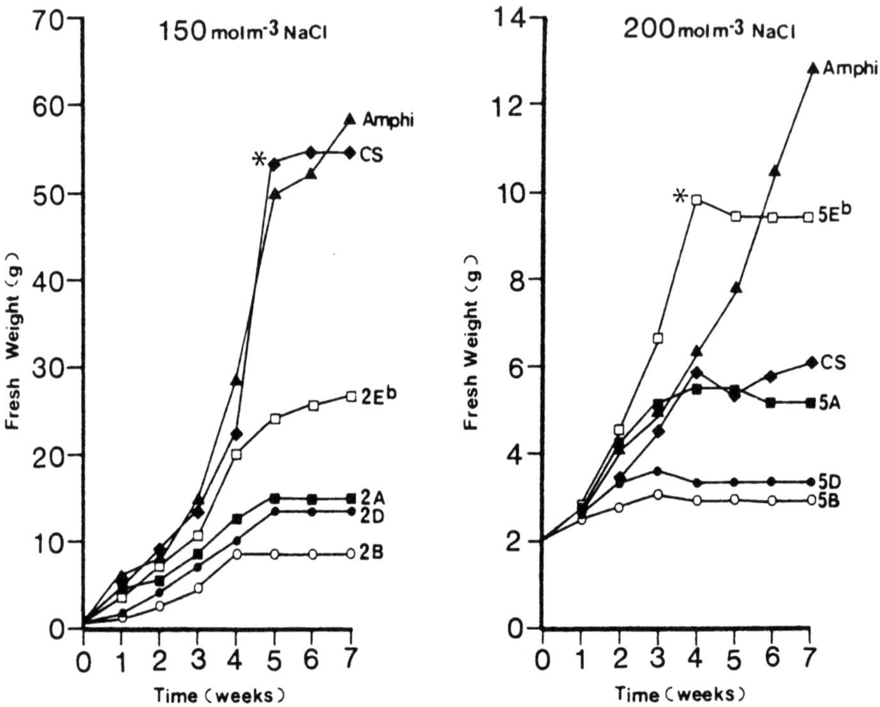

Fig. 4.4. Growth curves of wheat (CS), wheat/*Thinopyrum bessarabicum* amphiploid, $2E^b$ and $5E^b$ addition lines and relevant wheat tetrasomic lines grown at 150 mol m^{-3} and 200 mol m^{-3}. The amphiploid grew vegetatively throughout the timecourse of each experiment and produced exponential growth curves. At 150 mol m^{-3}, CS flowered (*) and produced a sigmoid growth curve typical of a completed life cycle, as did the tolerant $5E^b$ addition line at 200 mol m^{-3} NaCl. The plateau of other curves indicates plant death. Note also that production at 200 mol m^{-3} is about one-tenth of that at 150 mol m^{-3}

CIMMYT cultivar Glennson 81, a photoperiod-insensitive, 1BL/1RS semi-dwarf cultivar. Glennson 81 was therefore primed as a recipient for salt tolerance genes from *Th. bessarabicum.* This work is continuing at the time of writing. Some wheat/*Thinopyrum* translocations have been produced, but have not been tested (T. E. Miller pers. comm.).

A similar approach to understanding the genetic control of salt tolerance in the Triticeae has been pursued by Dvorak and his co-workers. *Thinopyrum elongatum* (syns: *Agropyron elongatum, Elytrigia elongata, Lophopyrum elongatum*) is a close relative of *Th. bessarabicum*; they are both diploid species, highly tolerant to salt, and have related genomes (Forster and Miller 1989). The E genome of *Th. elongatum* differs from the E^b genome of *Th. bessarabicum* by two translocations (Wang and Hsiao 1989). An octoploid amphiploid (2n = 8x = 56, AABBDDEE) between wheat and *Th. elongatum* was shown to have greater tolerance to salt than its wheat parent (Dvorak and Ross 1986). Like the wheat/*Th. bessarabicum* amphiploid, the wheat/*Th. elongatum* amphiploid is able to grow and set seed at 250 mol m^{-3} NaCl (Dvorak et al. 1988). Disomic, ditelosomic and chromosome substitution lines have been developed from this amphiploid (Dvorak and Sosulski 1974), and these have enabled *Th. elongatum* chromosomes involved in salt tolerance to be identified. Chromosomes 3E, 4E and 7E were found to have small but additive effects on tolerance to salt (Dvorak et al. 1988). These are not the same chromosome groups identified by Forster et al. (1988), i.e. $2E^b$ and $5E^b$, the disparity between the two sets of results may be due to the chromosome translocation differences between the two species.

4.5 The Development of Hybrids as New Crop Species

Hybrids between crops and their wild relatives are often produced as bridging species in crop improvement programmes. These hybrids often show an intermediary level of tolerance to abiotic stresses which diminishes with backcrossing to the crop. This is typical of characters which are controlled by several genes. One way around this problem is to develop hybrids as crop species in their own right. This approach is being pursued by Schachtman et al. (1991) in producing a salt-tolerant hexaploid wheat by hybridising *T. durum* (AABB) with salt-tolerant accessions of *Ae. squarrosa* (DD). The hybrid between *T. durum* and *Th. bessarabicum* also has potential to be developed into a new salt-tolerant crop species. These hybrids would be developed using a plant breeding programme similar to that of triticale, the wheat/rye hybrid (AABBRR). It took about 10 years to develop triticale into an important cereal crop (Gregory 1987).

4.6 The Interface Between Cytogenetics and Physiology

"With environmental stresses such as salinity or drought, a genetic approach that is divorced from an understanding of plant physiology is unlikely to succeed" (Munns et al. 1990). The same could be said for genetics and plant breeding. While physiological processes affected by salt stress are important to physiologists, plant survival and yield are of prime concern to the breeder. It is not for me, here, to pit one approach against the other, but there is no doubt that current research results are bringing the three disciplines of genetics, plant breeding and physiology closer together.

In some of the very first experiments of the cytogenetic programme at PBI, a link was established between a major physiological event, flowering time, and tolerance to salt. It was found that the early flowering cultivar Glennson 81 succumbed to salt damage (150 mol m^{-3} NaCl) earlier than the later flowering Chinese Spring. However, if Chinese Spring were vernalized, thus bringing forward the time to flowering, it, too, died earlier than the unvernalized controls.

In all cases, the time of death was close to the transition from vegetative to generative growth. This stage in development seemed to be particularly susceptible to salt damage, or else it was a stage when previously occurring metabolic damage became evident. The ability to flower at 150 mol m^{-3} NaCl was subsequently used at a test for salt tolerance. Tests based on the ability to complete the life cycle showed that the wheat/*Th. bessarabicum* amphiploid was highly tolerant to salt and that homoeologous group 2 and group 5 chromosomes have effects on tolerance (Forster et al. 1988). The effect of salt stress on various physiological characters has been studied in these lines (M. Mahmood pers. comm.). The 2E^{b} and 5E^{b} chromosome addition lines were found to accumulate more sodium in their leaves than the salt-tolerant amphiploid, but significantly less than that found in euploid wheat.

Homoeologous group 2 and group 5 chromosomes of the Triticeae are the sites of two major genes controlling days to flowering. *Ppd* genes (photoperiod insensitivity) and *Vrn* genes (reduced vernalization requirement) are located on group 2 (Scarth and Law 1983) and group 5 homologues (Law et al. 1975), respectively. It is possible that the response of group 2 and group 5 chromosomes addition lines to salt stress is due to the expression of these genes. Single chromosome recombinant lines for chromosomes 2B and 5A have made it possible to compare the effects of contrasting alleles of *Vrn1*/*vrn1* and *Ppd2*/*ppd2* in a uniform genetic background, thus minimising the effects of other genes (Taeb et al. 1993). The presence of a dominant allele at either locus was associated with lower average shoot Na concentration in both salt and non-stressed conditions. Since no interaction was found between genotype and salt treatment, it was suggested that tolerance genes are not switched on by stress, but instead may be active in non-stress conditions. The results indicate that *Vrn1* and *Ppd2* have pleiotropic effects on tolerance to salt. Another possibility is that these genes are tightly linked to some other tolerance genes. It is not known whether *Vrn* and *Ppd* genes are regulatory or structural, but since they respond to two different environmental stimuli, vernalization and day length, it is likely that they are regulatory, in which case the possibility of a single structural gene for tolerance to salt exists (Taeb et al. 1993).

There have been conflicting reports on whether salt stress activates specific genes. Ramagopal (1987) demonstrated differential transcription of mRNA in various barley cultivars when salt-stressed. However, Hurkman et al. (1989) studying the same cultivars could not show synthesis of any induced polypeptide or translatable mRNA, though salt stress did effect quantitative differences in mRNA synthesis. Gulick and Dvorak (1987) also could not find any novel mRNAs in the salt-tolerant wheat/*Th. elongatum* amphiploid when subjected to salt. The few results from molecular genetic work tend to favour the idea that salt stress does not activate genes for tolerance, but rather the innate genetic make-up of the plant determines fitness to salt stress.

4.7 Genetic Control of Salt Tolerance in Barley

Barley, *Hordeum vulgare*, is the most salt-tolerant of the major cereal crops. This has been known for some time, with dramatic effects on people and civilisations: barley completely replaced wheat in ancient Mesopotamia as irrigated lands became too saline for wheat production (Jacobsen and Adams 1958). Various tests and selections have been carried out in recent years on large and diverse populations of barley and significant and useful variation is available within the species for crop improvement (Storey and Wyn Jones 1978; Epstein et al. 1979; Jana et al. 1980; Omara et al. 1987; Ye et al. 1987).

The genes affecting tolerance to salt have been assigned to specific chromosomes of two barley species, *Hordeum vulgare* and *H. chilense* (Forster et al. 1990). Wheat/*H. vulgare* and wheat/*H. chilense* disomic addition lines were subjected to salt stress, and growth and yield were compared to control conditions. Plant vigour was found to have a major effect on tolerance to salt: vigorous genotypes in control conditions also performed well in saline conditions. Genes with positive effects on vigour were located on chromosomes 6H and 7H of *H. vulgare* and $6H^{ch}$ and $7H^{ch}$ of *H. chilense*. Specific chromosomes with direct effects on tolerance to salt were also identified, these were 4H and 5H of *H. vulgare* and $1H^{ch}$, $4H^{ch}$ and $5H^{ch}$ of *H. chilense*. It was also found that wheat/*H. chilense* addition lines performed better than those of wheat/*H. vulgare*. The wild species, therefore, has more potent genes for tolerance to salt than the barley cultivar Betzes used to produce the wheat/*H. vulgare* additions (Islam et al. 1981). This emphasises the point that wild species are useful sources of novel genes. Fertile hybrids between *H. vulgare* and *H. chilense* have not been produced and introgression of genes between the two species is not possible through the normal route of meiotic chromosome recombination.

There are, however, other genetic strategies open for improving the salt tolerance of barley. Genes for both tolerance to salt and vigour are important in this context. These genes are located on different and potentially opposing chromosomes; group 6 and 7 chromosomes carry genes for vigour, but these chromosome addition lines have the worst response to salt stress relative to control conditions; homoeologous group 4 and 5 chromosomes carry genes for tolerance to salt, but can also confer poor vigour. There is, therefore, a dilemma. One option is to select on the basis of vigour. Salt stress is known to reduce growth, and vigorous lines would be expected to have an advantage: vigorous lines should, therefore, yield well in both fertile and saline conditions (a strategy supported by Richards 1983). In this instance, genes on group 6 and 7 chromosomes would be of importance. Alternatively, genes with a direct positive effect can be exploited. Homoeologous group 4 and 5 chromosomes of salt-tolerant wild barley *H. spontaneum* probably offer the best source of potent salt tolerance genes which can be introduced easily into the crop species, *H. vulgare*, by hybridisation and backcrossing.

Table 4.1. Chromosomal location of genes for mineral stress tolerance in the Triticeae

Mineral stress	Species	Chromosome (S = short arm, L = long arm)							Reference[a]
Aluminium	*Secale cereale*			3R	4R	5R	6RS		1, 2
	Secale montanum		$2R^{m}$			$5R^{m}S$			2
	Triticum aestivum						6AL	7AS	2
					$4A^{b}S$				2
			2DL	3DL	4DL	5D		7D	2, 3, 4
Boron	*Aegilops sharonensis*			$3S^{l}$		$5S^{l}$		$7S^{l}$	5
	Secale cereale		2R	3R		5R	6R	7RL	5
	Thinopyrum elongatum	1E	2E	3E		5E	6E	7E	5
	Triticum aestivum				$4A^{b}$				6
Copper	*Secale cereale*		2R						5
	Secale montanum		$2R^{m}S$						5
	Thinopyrum bessarabicum		$2E^{b}$						5
	Triticum aestivum					5A		7A	5
	Triticum spelta					5A		7B	5
					4D			7D	5
Manganese	*Secale cereale*			3R			6R		5
	Thinopyrum bessarabicum					$5E^{b}$			5
Sodium	*Hordeum vulgare*				4H	5H			7
	Hordeum chilense	$1H^{ch}$			$4H^{ch}$	$5H^{ch}$			7
	Thinopyrum bessarabicum		$2E^{b}$			$5E^{b}$			8
	Thinopyrum elongatum			3E	4E			7E	9
	Triticum aestivum				4D				10

[a] Reference: 1, Aniol and Gustafson (1984); 2, Manyowa et al. (1988); 3, Polle et al. (1978); 4, Prestes et al. (1975); 5, Manyowa (1989); 6, Paull et al. (1988); 7, Forster et al. (1990); 8, Forster et al. (1988); 9, Dvorak et al. (1988); 10, Gorham et al. (1987).
[b] Previously designated 4B.

Salt-tolerant populations of *H. spontaneum* have been collected from various locations in Israel. These have been crossed onto a *H. vulgare* cultivar as a first step in genetic studies to determine the dominance and location of salt tolerance genes and transfer them into the crop from wild relatives. Barley probably has the greatest potential of becoming the first truly salt-tolerant grain crop in the near future.

4.8 Genes for Abiotic Stress Tolerance in the Triticeae

The chromosome location of genes controlling tolerance to several mineral stresses is given in Table 4.1. The results have been compiled from various reports which have exploited aneuploid genetic stocks to locate tolerance genes to specific chromosomes. It is clear that genes controlling tolerance are located on all seven homoeologous chromosome groups of the Triticeae. However, group 4 and group 5 chromosomes are predominant and are involved in all five stresses listed. It is not known whether this multi-tolerance is due to the expression of the same gene(s) or that these chromosome families carry gene clusters for abiotic stress tolerance. Although mineral stress tolerance is influenced by genes dispersed over entire genomes, single genes with major effects are known. Macnair (1983) reported a single major gene controlling tolerance to copper in yellow monkey flower, *Mimulus guttatus* (see also Chap. 7). In soybean, *Glycine max*, chloride exclusion is controlled by the single gene, *Ncl* (Abel and MacKenzie 1964). Single genes for tolerance to aluminium have also been reported in maize, *Zea mays* (Rhue et al. 1978) and in barley, *Hordeum vulgare* (Reid 1970). Gene clusters of stress tolerance genes have been reported in plasmids of bacteria, and in plants such as lettuce and maize (see Devine 1982 for a review) and this may indicate that the clustering of genes with similar functions is a universal phenomenon. If tolerance genes are tightly clustered, their manipulation by chromosome engineering techniques would be the same as that described above for introducing small chromosome segments.

4.9 Examples of Alien Introduction

Crop species often lack characters which can be found in related species, and it would be advantageous if genes controlling these traits could be introgressed into the crop. Wide hybridisation has proved very useful in the transfer of alien genes into wheat. Where a common genome exists between the wild species and the crop, gene transfers can be relatively easy, since there is no barrier to meiotic recombination. Notable examples of such homologous gene transfer into wheat include resistance to powdery mildew from *T. timopheevi* (AAGG) (Allard and Shands 1954) and resistance to eyespot from *Aegilops ventricosa* (M^vM^vDD)

(Maia 1967). Gene transfer from relatives without a common genome requires more work. Techniques in gene transfer have become progressively more sophisticated. Early work concentrated on producing hybrids and amphiploids in which the whole genome of a wild species was added to wheat. Apart from triticale, the wheat/rye hybrid, most of these hybrids were unsuccessful since they possessed poor agronomic characters. The effects of deleterious genes can be reduced by the production of chromosome addition, substitution and translocation lines. Translocations involving chromosomes of rye and *Th. elongatum* carrying disease resistance genes have been cited above; but these lines still carry a relatively large amount of alien DNA with negative as well as positive effects. Work is now concentrated on the transfer of small segments of chromosomes. This can be achieved in some polyploid species by the suppression of homoeologous pairing: in wheat this can be done using genetic stocks lacking *Ph1*, but other techniques are available. In the cultivated oat, *Avena sativa*, bivalent formation can be modified by the effect of genes of the diploid species *A. longiglumis* (Rajhathy and Thomas 1972). The effect of *A. longiglumis* has been used to transfer mildew resistance from *A. barbata* to the crop, *A. sativa*. In wheat, the activity of *Ph1* can be suppressed by the related species *Ae. speltoides* and *Ae. mutica* (Riley et al. 1968) and this has been used to transfer mildew resistance from *Ae. speltoides* to wheat. The manipulation of *Ph1* by either suppression, absence or mutation is the most promising means of alien transfer in wheat. The absence of an active *Ph1* does not, however, guarantee homoeologous recombination. The R genome of rye and the H genome of barley are relatively unresponsive to this type of manipulation, although Koebner and Shepherd (1986) have shown 1.5% recombination between 1RL and wheat homoeologues using the *ph1b* mutant (Sears 1977). Larkin et al. (1990) have put forward an alternative strategy. Cell cultures of wheat/alien chromosome addition lines have been used successfully to transfer into wheat genes for resistance to cereal cyst nematode from rye and resistance to barley yellow dwarf virus from *Th. intermedium*. It has been suggested that the technique can be used to transfer chromosome segments into crops which have genetic barriers to meiotic recombination. Other methods of producing chromosome translocations have included the use of radiation. Sears (1956) was the first to use irradiation-induced translocation to transfer brown rust resistance from *Ae. umbellulata* to wheat. One advantage of introducing alien chromosome segments is that the introduction is fixed in the presence of the wild-type *Ph1* gene. The possibility of introducing several genes in this way has been proposed by Law et al. (1983), a concept known as the super gene. For example, the *Ae. comosa* segment in Compair wheat carries genes for both resistance to stripe rust (Riley et al. 1968) and increased grain protein, and this segment does not naturally recombine with wheat (Law et al. 1984). A comprehensive catalogue of alien introductions in wheat can be found in Gale and Miller (1987).

4.10 Genetic Markers in Plant Breeding

Genetic studies in the Triticeae have involved mapping genes to chromosomes. The genetic maps are based largely on morphological, protein, disease resistance and restriction fragment length polymorphism (RFLP) loci (see McIntosh 1988 for wheat, von Wettstein-Knowles 1992 for barley). As the density and detail of these maps increases, linkage between genes of agronomic importance and easily detected loci can be established. The latter can then be used as genetic markers for the indirect selection of the desired trait. Conventional markers include morphological traits, such as the hairy neck character used to detect the presence of chromosome 5RL of rye (Sears 1967), and cytological markers such as C-banding and in situ hybridisation which can detect specific chromosomes or chromosome segments (Gill et al. 1988). Techniques in biochemistry and molecular genetics have produced other markers. For example, *Pch1*, a major gene for resistance to eyespot disease of wheat, can be selected by its close linkage with the endopeptidase gene *Ep-1* using isozyme analysis (Worland et al. 1988). Similarly, RFLP analysis in barley can be used to detect a major gene, *Sh*,

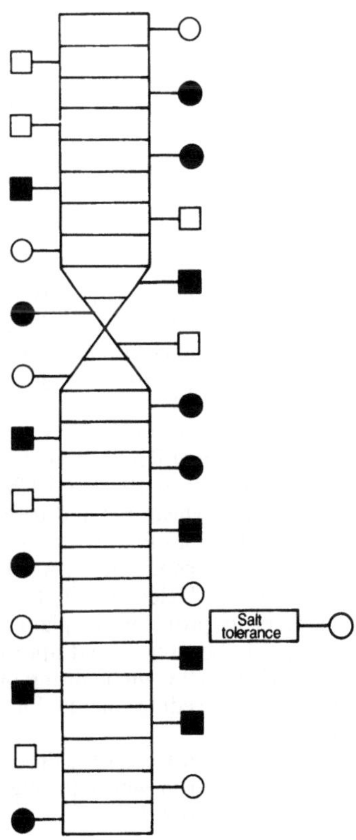

Fig. 4.5. A hypothetical chromosome with various genetic markers (cytological, morphological, isozyme, RFLP etc.) along its entire length

determining spring/winter habit (Chojecki et al. 1989). A new generation of genetic markers is emerging which can detect polymorphism using DNA amplification by the polymerase chain reaction (D'Ovidio et al. 1990). The use of genetic markers for salt tolerance and other abiotic stress genes will greatly increase the speed and precision of gene transfer, and such markers are actively being sought (Fig. 4.5).

4.11 Summary

Crop improvement for stress tolerance has been traditionally centred on the disciplines of physiology, biochemistry and plant breeding. The genetic approach has emerged fairly recently and offers great potential. Genes controlling tolerance to various mineral stresses have been located to specific chromosomes and potent alleles of these genes can be found in wild and related species in the Triticeae. A battery of techniques for alien gene transfer are now available and successful transfers have been developed, the end product being commercial cultivars. The ability of the cytogeneticist to remodel the chromosomes of crop species is having a major impact and will continue to play a key role in future crop improvement goals.

Acknowledgements. I would like to thank Drs N. Clipson, N. M. Manyowa, R. Munns, W. Powell, G. Ramsay, M. Taeb for their advice and criticism of this chapter. Thanks are also due to Mr I. Morrison for artwork.

References

Abel GH, Mackenzie AJ (1964) Salt tolerance of soybean varieties during germination and later growth. Crop Sci 4: 157

Allard RW, Shands RG (1954) Inheritance of resistance to stem rust and powdery mildew in cytologically stable spring wheats derived from *Triticum timopheevi.* Phytopathology 44: 266–274

Alonso LC, Kimber G (1980) A hybrid between diploid *Agropyron junceum* and *Triticum aestivum.* Cereal Res Commun 8: 355–358

Aniol A, Gustafson PJ (1984) Chromosome location of genes controlling aluminium tolerance in wheat, rye and triticale. Can J Cytol 26: 701–705

Charnock A (1988) Plants with a taste for salt. New Sci 3rd Dec: 41–45

Chojecki J, Barnes S, Dunlop A (1989) A molecular marker for vernalisation requirement in barley. In: Helentjaris T, Burr B (eds) Development and application of molecular markers to problems in plant genetics. Current Commun Mol Biol, Cold Spring Harbor Laboratory, Cold Spring Harbor, pp 145–148

Devine TE (1982) Genetic fitting of crops to problem soils. In: Christiansen MN, Lewis CF (eds) Breeding plants for less favourable environments. Wiley, New York, pp 143–158

D'Ovidio R, Tanzarella OA, Proceddu E (1990) Rapid and efficient detection of genetic polymorphism in wheat through amplification by polymerase chain reaction. Plant Mol Biol 15: 169–171

Dvorak J, Knott DR (1973) Disomic and ditelosomic additions of diploid *Agropyron elongatum* chromosomes to *Triticum aestivum*. Can J Genet Cytol 16: 399–417
Dvorak J, Ross K (1986) Expression of tolerance of Na^+, K^+, Mg^+, Cl^- and SO_4^{2-} and sea water in the amphiploid of *Triticum aestivum* × *Elytrigia elongata*. Crop Sci 26: 658–660
Dvorak J, Sosulski FW (1974) Effect of additions and substitutions of *Agropyron elongatum* chromosomes on quantitative characters in wheat. Can J Genet Cytol 16: 627–637
Dvorak J, Edge M, Ross K (1988) On the evolution of the adaptation of *Lophopyrum elongatum* to growth in saline environments. Proc Natl Acad Sci USA 85: 3805–3809
Epstein E, Kingsbury RW, Norlyn JD, Rush DW (1979) An approach to utilisation of underexploited resources. In: Hollaender A (ed) The biosaline concept. Plenum Press, New York, pp 77–79
Farooq S, Niazi MLK, Iqbal N, Shah TM (1989) Salt tolerance potential of wild resources of the tribe Triticeae II. Screening of species of *Aegilops*. Plant Soil 119: 255–260
Forster BP, Miller TE (1985) A 5B-deficient hybrid between *Triticum aestivum* and *Agropyron junceum*. Cereal Res Commun 13: 93–95
Forster BP, Miller TE (1989) Genome relationship between *Thinopyrum bessarabicum* and *Thinopyrum elongatum*. Genome 32: 930–931
Forster BP, Gorham J, Miller TE (1987) Salt tolerance of an amphiploid between *Triticum aestivum* and *Agropyron junceum*. Plant Breed 98: 1–8
Forster BP, Miller TE, Law CN (1988) Salt tolerance of two wheat/*Agropyron junceum* disomic addition lines. Genome 30: 559–564
Forster BP, Phillips MS, Miller TE, Baird E, Powell W (1990) Chromosome location of genes controlling tolerance to salt (NaCl) and vigour in *Hordeum vulgare* and *H. chilense*. Heredity 65: 99–107
Gale MD, Miller TE (1987) The introduction of alien genetic variation into wheat. In: Lupton FGH (ed) Wheat breeding, its scientific basis. Chapman and Hall, London, pp 173–210
Gill BS, Lu F, Schlegel R, Endo TR (1988) Toward a cytogenetic and molecular description of wheat chromosomes. In: Miller TE, Koebner RMD (eds) Proc 7th Int Wheat Genet Symp. Cambridge University Press, Cambridge UK, pp 477–481
Gorham J, Budrewicz E, Wyn Jones RG (1985) Salt tolerance in the Triticeae: growth and solute accumulation in leaves of *Thinopyrum bessarabicum*. J Exp Bot 36: 1021–1031
Gorham J, Forster BP, Budrewicz E, Wyn Jones RG, Miller TE, Law CN (1986a) Salt tolerance in the Triticeae: salt accumulation and distribution in an amphiploid derived from *Triticum aestivum* cv. Chinese Spring and *Thinopyrum bessarabicum*. J Exp Bot 37: 1435–1449
Gorham J, Budrewicz E, McDonnell E, Wyn Jones RG (1986b) Salt tolerance in the Triticeae: salinity-induced changes in the leaf solute composition of some perennial Triticeae. J Exp Bot 37: 1114–1128
Gorham J, Hardy C, Wyn Jones RG, Joppa LR, Law CN (1987) Chromosome location of a K/Na discrimination character in the D genome of wheat. Theor Appl Genet 74: 584–588
Greenway H, Munns R (1980) Mechanism of salt tolerance in nonhalophytes. Annu Rev Plant Physiol 31: 149–190
Gregory RS (1987) Triticale breeding. In: Lupton FGH (ed) Wheat breeding, its scientific basis. Chapman and Hall, London, pp 269–286
Gulick P, Dvorak J (1987) Gene induction and repression by salt treatment in roots of salinity-sensitive Chinese Spring wheat and the salinity tolerant Chinese Spring × *Elytrigia elongata* amphiploid. Proc Natl Acad Sci USA 74: 99–103
Hurkman WJ, Fornari CS, Tanaka CK (1989) A comparison of the effect of salt on polypeptides and translatable mRNAs in roots of a salt-tolerant and a salt-sensitive cultivar of barley. Plant Physiol 90: 1444–1456

Islam AKMR, Shepherd KW, Sparrow DHB (1981) Isolation and characterisation of euplasmic wheat-barley chromosome addition lines. Heredity 46: 161–174
Jacobsen T, Adams RM (1958) Salt and silt in ancient Mesopotamian agriculture. Science 128: 1251–1258
Jana MK, Jana S, Acharya SN (1980) Salt tolerance in heterogeneous populations of barley. Euphytica 29: 409–417
Joshi YC, Qadar A, Rana RS (1979) Differential sodium and potassium accumulation related to sodicity tolerance in wheat. Indian J Plant Physiol 22: 226–230
Kingsbury RW, Epstein E (1984) Selection for salt-resistant spring wheat. Crop Sci 24: 310–315
Knott DR (1968) Agropyrons as a source of rust resistance in wheat breeding. In: Finlay KW, Shepherd KW (eds) Proc 3rd Int Wheat Genet Symp, Canberra, Australia, pp 204–212
Koebner RMD, Shepherd KW (1986) Controlled introgression to wheat of genes from rye chromosome 1RS by induction of allosyndesis I. Isolation of recombinants. Theor Appl Genet 73: 197–208
Larkin PJ, Spindler LH, Banks PM (1990) Cell culture of alien chromosome addition lines to induce somatic recombination and gene introgression. In: Nijkamp HJJ, van der Plas LHW, van Aartrijk J (eds) Progress in plant cellular and molecular biology. Kluwer, Dordrecht, pp 163–168
Law CN, Worland AJ, Giorgi B (1975) The genetic control of ear emergence time by chromosomes 5A and 5D of wheat. Heredity 36: 49–58
Law CN, Snape JW, Worland AJ (1983) Chromosome manipulation and its exploitation in genetics and breeding of wheat. Proc Stadler Symp, University of Columbia, MO, 5: 5–23
Law CN, Payne PI, Worland AJ, Miller TE, Harris PA, Snape JW, Reader SM (1984) Cereal grain protein improvement. Int Atomic Energy Agency, Vienna, pp 279–300
Law CN, Snape JW, Worland AJ (1987) Aneuploidy in wheat and its uses in genetic analysis. In: Lupton FGH (ed) Wheat breeding, its scientific basis. Chapman and Hall, London, pp 71–107
Maan SS, Gordon J (1988) Compendium of alloplasmic lines and amphiploids in the Triticeae. In: Miller TE, Koebner RMD (eds) Proc 7th Int Wheat Genet Symp. Cambridge University Press, Cambridge, pp 1325–1371
Macnair MR (1983) The genetic control of copper tolerance in the yellow monkey flower, *Mimulus guttatus*. Heredity 50: 283–293
Maia N (1967) Obtention de blés tendres resistants au pietin-verse (*Cercosporella herpotrichoides*) par croisements interspecifiques. CR Seances Acad Agric Fr 53: 149–154
Manyowa NM (1989) The genetics of aluminium, excess boron, copper and manganese stress tolerance in the tribe Triticeae and its implication for wheat improvement. PhD Thesis, University of Cambridge, Cambridge
Manyowa NM, Miller TE, Forster BP (1988) Alien species as sources for aluminium tolerance genes for wheat (*Triticum aestivum*). In: Miller TE, Koebner RMD (eds) Proc 7th Int Wheat Genet Symp. Cambridge University Press, Cambridge, pp 851–857
McIntosh RA (1988) Catalogue of gene symbols for wheat. In: Miller TE, Koebner RMD (eds) Proc 7th Int Wheat Genet Symp. Cambridge University Press, Cambridge, pp 1225–1324
Mettin D, Bluethner WD, Weinrich M (1973) Studies on the nature and the possible origin of the spontaneously translocated 1B-1R chromosome in wheat. Wheat Info Serv 47/48: 12–16
Miller TE (1987) Systematics and evolution. In: Lupton FGH (ed) Wheat breeding, its scientific basis. Chapman and Hall, London, pp 1–30
Munns R, Schachtman DP, Lagudah ES, Appels R (1990) Improving salt tolerance in wheat – a combined physiological and genetic approach. Proc Int Symp Molecular and Genetic Approaches to Plant Stress. ICGEB, New Delhi, pp 30.1–30.2

Okamoto M (1957) Asynaptic effect of chromosome V. Wheat Info Serv 5: 6
Omara MK, Abel-Rahman KA, Hussein MY (1987) Selection for salt stress tolerance in barley. Assiut J Agric Sci 18: 199–218
Paull JG, Rathjen AJ, Cartwright B (1988) Genetic control of tolerance to high concentrations of boron in wheat. In: Miller TE, Koebner RMD (eds) Proc 7th Int Wheat Genet Symp. Cambridge University Press, Cambridge, pp 871–877
Pearce F (1987) Banishing the salt of the earth. New Sci 11th June: 53–56
Polle E, Konzak CF, Kittrick JA (1978) Rapid screening of wheat for tolerance to Al in breeding varieties better adapted to acid soils. Agricultural Technology for Developing Countries. Tech Ser Bull 21 AID, Washington DC
Prestes AM, Konzak CF, Kittrick JA (1975) An improved seedling culture method for screening wheat for tolerance to toxic levels of aluminium. Agro Abstr 67: 60
Quershi RH, Ahmad R, Ilyas M, Aslam Z (1980) Screening wheat (*Triticum aestivum* L.) for salt tolerance. Pak J Agric Sci 17: 19–25
Rajaram S, Maan CLE, Ortiz-Ferrara G, Mujeeb-Kazi A (1983) Adaptation, stability and high yield potential of certain 1B/1R CIMMYT wheats. In: Saskamoto S (ed) Proc 6th Int Wheat Genet Symp, Kyoto, Japan, pp 613–621
Rajhathy T, Thomas H (1972) Genetic control of chromosome pairing in hexaploid oats. Nature New Biol 239: 217–219
Ramagopal S (1987) Differential mRNA transcription during salinity stress in barley. Proc Natl Acad Sci USA 84: 94–98
Rana RS (1986) Evaluation and utilisation of traditionally grown cereal cultivars of salt affected areas in India. Indian J Genet 46(Suppl 1): 121–135
Reid DA (1970) Genetic control of reaction to Al in winter barley. In: Nilan RA (ed) Proc 2nd Int Barley Genet Symp. Washington State University Press, Washington, pp 409–413
Rhue RD, Grogan CD, Stockmeyer EW, Everett HL (1978) Genetic control of aluminium tolerance in corn. Crop Sci 18: 1063–1067
Richards RA (1983) Should selection for yield in saline regions be made on saline or non-saline soils? Euphytica 32: 431–438
Riley R, Chapman V, Johnson R (1968) The incorporation of disease resistance in wheat by genetic interference with the regulation of meiotic chromosome synapsis. Genet Res 12: 199–219
Sayed J (1985) Diversity of salt tolerance in a germplasm collection of wheat (*Triticum aestivum*). Theor Appl Genet 69: 651–657
Scarth R, Law CN (1983) The location of the photoperiodic genes *Ppd2* and an additional factor for ear emergence time on chromosome 2B of wheat. Heredity 51: 607–619
Schachtman DP, Munns R, Whitecross MI (1991) Variation in Na exclusion and salt tolerance in *Triticum tauschii* (Coss) Schmal. Crop Sci 31: 992–997
Schulz-Schaeffer J (1980) History of cytogenetics. In: Schulz-Schaeffer J (ed) Cytogenetics. Springer, Berlin Heidelberg New York, pp 2–29
Sears ER (1956) The transfer of leaf-rust resistance from *Aegilops umbellulata* to wheat. In: Genetics in plant breeding. Brookhaven Symp Biol 2: 1–12
Sears ER (1967) Induced transfer of hairy neck from rye to wheat. Z Pflanzenzucht 57: 4–25
Sears ER (1973) Translocation through union of newly formed telocentric chromosomes. Proc 13th Int Congr Genet, Berkeley, CA. Genetics (Suppl 1) 2: 247
Sears ER (1977) An induced mutant with homoeologous pairing in common wheat. Can J Genet Cytol 19: 585–593
Shah SH, Gorham J, Forster BP, Wyn Jones RG (1987) Salt tolerance in the Triticeae: the contribution of the D genome to cation selectivity in hexaploid wheat. J Exp Bot 38: 254–269
Shepherd KW, Islam AKMR (1988) Fourth compendium of wheat-alien chromosome lines. In: Miller TE, Koebner RMD (eds) Proc 7th Int Wheat Genet Symp. Cambridge University Press, Cambridge, pp 1373–1397

Stoner R (1988) Engineering a solution to the problem of salt-laden soils. New Sci 3 Dec: 44

Storey R, Wyn Jones RG (1978) Salt stress and comparative physiology in the Gramineae. I. Ion relations in two salt-and water stressed barley cultivars, California Mariout and Arimar. Aust J Plant Physiol 5: 801–816

Taeb M, Koebner RMD, Forster BP (1993) Adaptive effects of genes for vernalization and photoperiod requirement on salinity tolerance of wheat (*Triticum aestivum* L.). Plant Cell Environ (in press)

Von Wettstein-Knowles P (1992) Cloned and mapped genes: current status. In: Shewry PR (ed) Barley: genetics, biochemistry, molecular biology and biotechnology. CAB International, Wallingford, pp 73–98

Wang RR-C, Hsiao C (1989) Genome relationship between *Thinopyrum bessarabicum*, and *Thinopyrum elongatum*: revisited. Genome 32: 802–809

Worland AJ (1988) Catalogue of monosomic series. In: Miller TE, Koebner RMD (eds) Proc 7th Int Wheat Genet Symp. Cambridge University Press, Cambridge, pp 1388–1403

Worland AJ, Law CN, Hollins TW, Koebner RMD, Giura A (1988) Location of a gene for resistance to eyespot (*Pseudocercosporella herpotrichoides*) on chromosome 7D of bread wheat. Plant Breed 101: 43–51

Wyn Jones RG, Gorham J, McDonnell E (1984) Organic and inorganic solute contents as selection criteria for salt tolerance in the Triticeae. In: Staples RC, Toenniessen GH (eds) Salinity tolerance in plants. Wiley, New York, pp 189–203

Ye JM, Kao KN, Harvey BL, Rossnagel BG (1987) Screening salt-tolerant barley genotypes via F_1 anther culture in salt stress media. Theor Appl Genet 74: 426–429

Zeller FJ (1973) 1B/1R Wheat-rye chromosome substitutions and translocations. In: Sears ER, Sears LM (eds) Proc 4th Int Wheat Genet Symp, University of Columbia, MO, pp 209–221

Chapter 5

Tissue Culture in the Improvement of Salt Tolerance in Plants

P. M. HASEGAWA, R. A. BRESSAN, D. E. NELSON, Y. SAMARAS, and D. RHODES

5.1 Introduction

The use of tissue culture for improving osmotic (water and salt) stress tolerance in plants has been envisaged primarily on the assumed agricultural potential that will arise from the unique genetic manipulations that can be imposed on cells and tissues in vitro (Nabors 1976; Stavarek and Rains 1984; Raghava Ram and Nabors 1985; Spiegel-Roy and Ben-Hayyim 1985; Hasegawa et al. 1986; Tal 1990). Plant scientists who are predisposed to this philosophy, and not all are, typically view the application of tissue culture from the perspective of how it could augment traditional breeding approaches directed towards the improvement of salinity tolerance in plants. This includes the utilisation of techniques, such as embryo or ovule rescue, that would substantially support efforts to introgress genetic determinants of stress tolerance through wide crosses and interspecific hybridisation. However, substantially greater utility of tissue culture to achieve this goal is hypothesised as a result of potential to exploit effectively sources of useful genetic variation through highly efficient cell or tissue selection strategies, parasexual hybridisation and gene transformation.

In parallel with these more direct tissue culture applications, we and others have proposed and, in our view, demonstrated the potential utility of in vitro systems in research to identify and characterise cellular determinants that are involved in salt stress tolerance (Heyser and Nabors 1981a, b; Watad et al. 1983; Harms and Oertli 1985; Stavarek and Rains 1985; Hasegawa et al. 1986, 1990; Ben-Hayyim et al. 1987; Binzel et al. 1989; Rhodes and Handa 1989; Singh et al. 1989; Bressan et al. 1990; Tal 1990; Dracup 1991; Schnapp et al. 1991). This approach is rationalised because of the scant amount of information that is available pertaining to the precise physiological, biochemical and molecular genetic mechanisms that causally effect adaptation to and growth in saline environments. Even less data are available about stress perception, transduction of signals to initiate and co-ordinate adaptive processes and the co-ordination of metabolism required in the new homeostatic steady state for growth in the saline environment. Information of this type is a prerequisite for the establishment of efficient selection criteria that would substantially assist traditional and non-traditional approaches to the introgression of salt tolerance characters into plants of agricultural significance.

Monographs on Theoretical and Applied Genetics, Vol. 21
Ed. by A. R. Yeo and T. J. Flowers

The advantages of cell and tissue cultures in studies to identify and characterise determinants of salt tolerance are probably best illustrated by drawing analogy to the rationale that prompted the use of in vitro systems in animal research. Technical benefit results from the capacity to impose very stringent nutritional and environmental controls on the experimental system (Binzel et al. 1985; Hasegawa et al. 1986). The reductionist approach substantially simplifies the entity or process to be dissected and takes on additional significance if the biological entity or process is inherently cellularly based. Often, the complexities of an organism make definitive mechanistic characterisation in vivo virtually impossible. It has been possible to establish genetic systems whereby highly salt-tolerant cells can be isolated from cell populations that are inherently very vulnerable to the stress (Bressan et al. 1985, 1987; Watad et al. 1985, 1991b; Tal 1990). Although not always viewed as mutants, these cells nonetheless have provided the opportunity to characterise putative mechanistic determinants in cells that are essentially isogenic for salt tolerance. The differentiation capacity of cells and tissues in vitro of many plants offers the potential to assess the contribution of cellular mechanisms for organismal adaptation and growth in salt.

In the remaining portions of this chapter, we will summarise the current status of research (Circa 1991) utilising tissue cultures for the purpose of improving salt tolerance in plants and provide an opinion about current prospects. Focus will be on somatic cell selection, parasexual hybridisation and gene transfer techniques and the utility of in vitro systems in research to identify and characterise mechanistic processes involved in salt tolerance. Although tissue cultures represent a subset of the integrated complexity of the organism, valuable information pertaining to fundamental cellular processes has been obtained that can be extrapolated to plants. In this context we discuss a proposed interrelationship between osmotic adjustment and cell expansion and its significance to adaptation to and growth in saline environments.

5.2 Application of Tissue Culture to Obtain Salt-Tolerant Plants

5.2.1 Somatic Cell Selection

Recent reviews (Tal 1990; Dracup 1991) provide a background summary of research that has been directed towards the goal of utilising somaclonal variation (variation amongst plants regenerated in vitro; Larkin and Scowcroft 1981) to obtain salt-tolerant plants. In the majority of plant species, considerable chromosomal instability occurs in populations of somatic cells (Partanen 1965; Torrey 1967; D'Amato 1977, 1990). This variability is confined primarily to differentiated cells; with meristems, including the progenitors of germ cells, exhibiting little instability. The conservation of genetic integrity in the latter cell types is critical to the preservation of the germplasm base. The tendency of meristematic cells to maintain genetic homogeneity appears to be inherent in their developmental programme as unorganised growth of these cells in vitro results in a substantial increase in genetic variation (Bayliss 1980; D'Amato 1985; Lee and Phillips 1988). Some somaclonal variation has been attributed to specific genetic alterations that occur in tissue culture (Larkin and Scowcroft 1981; Lee and Phillips 1988; Phillips et al. 1990; Brettell and Dennis 1991; Karp 1991).

Exploitation of genetic variability that exists in somatic cell populations has been restricted agriculturally to plants that are vegetatively propagated, i.e. principally horticultural species (Broertjes and van Harten 1978; Skirvin 1978). However, the pioneering work that defined the hormonal bases for the regulation of adventitious organogenesis (Skoog and Miller 1957) and embryogenesis (Reinert 1958a, b 1959; Steward et al. 1958) and the recent application of these concepts to numerous crop species, including graminaceous crops (Vasil 1987; Gobel and Lorz 1988; Bhaskaran and Smith 1990; Potrykus 1990), has substantially enhanced the agricultural potential for exploitation of somaclonal variation. Since genetic characterisation of somatic cell variability is still rather incomplete (Lee and Phillips 1988; Phillips et al. 1990; Karp 1991), it remains to be resolved whether it represents a unique potential source of variability for salt tolerance. Further, it is unclear how effectively this variability can be manipulated for agricultural utilisation. It is possible that this latter goal will be realised as a result of the development of highly efficient in vitro selection strategies even if genetic determinants of salt tolerance from somaclonal variation are analogous to those that are accessible through traditional breeding strategies. Since salt tolerance is generally viewed as being complex mechanistically, effective selection protocols should also be highly advantageous in schemes to pyramid multiple determinants into a single genotype.

There is little question that the frequency of genetic variation that occurs in somatic cells is extremely high (Bayliss 1980; D'Amato 1985; Lee and Phillips 1988; Tal 1990; Karp 1991). In fact, this frequency is often so high that there is question about how effectively it can be exploited to isolate genotypes whose genetic constitution is sufficiently unchanged, except for the addition of a new character of agricultural utility (Cai et al. 1990). If substantial linked or unlinked pleiotropic genetic anomalies accompany the selection for the desired trait, stabilising it in the appropriate genetic background may be no more efficient than the process of incorporating salt tolerance traits by interspecific hybridisation.

Research has provided examples that relative salt tolerance exhibited by a genotype as a plant could be correlated with that of its isolated cells and tissues in vitro (Orton 1980; Smith and McComb 1981a, b; Warren and Gould 1982; Tal 1984). However, this correlation was not always absolute (Strogonov 1973; Gale and Boll 1978; Smith and McComb 1981a; Hanson 1984; Flowers et al. 1985; McCoy 1987a). The lack of corroboration between plant and tissue culture data is attributable, in some instances, to the fact that certain plants utilise mechanisms that enhance their salt tolerance which are not based solely on the intrinsic capacities of actively growing cells. However, this cannot be the sole explanation for the discrepancies of enhanced salt tolerance expressed by cells in vitro relative to plants (Gale and Boll 1978; Flowers et al. 1985). We have determined that seemingly contradictory results pertaining to in vitro and in vivo responses of plants may, in some instances, depend upon the criterion used to evaluate salt tolerance (Casas et al. 1991). Although cells of the halophyte *Atriplex nummularia* do not exhibit in vitro a growth optimum at higher levels of NaCl than cells of the glycophyte tobacco, evaluation of the capacity to survive exposure to salt clearly delineates *A. nummularia* as a halophyte.

From the results of initial research, it is clear that plant cells capable of tolerating salt to a greater extent than corresponding wild-type cells can be readily isolated (Zenk 1974; Dix and Street 1975; Nabors et al. 1975; Goldner et al. 1977; Croughan et al. 1978; Hasegawa et al. 1980; Tyagi et al. 1981; Ben-Hayyim and Kochba 1982; Kochba et al. 1982). However, there were instances when a relative increase in salt tolerance manifested by cells or tissues isolated and maintained in the presence of salt was rapidly reduced upon transferral to media without the stress agent (Dix and Street 1975; Hasegawa et al. 1980; Chandler and Vasil 1984; Bressan et al. 1985). It is probable that the observed instability was, in some instances, the result of salt tolerance measurement inaccuracies resulting from phenomena such as cell morphometry, growth stage and density and physiological status of the cells (Watad et al. 1983, 1985; Binzel et al. 1985; Flowers et al. 1985; Harms and Oertli 1985; Hasegawa et al. 1986; Tal 1990; Dracup 1991).

However, another explanation is based on the capacity of isolated plant cells and tissues to adjust or acclimate to saline environments (Bressan et al. 1985; Hasegawa et al. 1986). Manifestation of this capacity is evident when cells are exposed to incremental increases in salt over sufficient time periods for adjustment to occur without lethality due to osmotic or ionic shock (Amar and Reinhold 1973). This attribute is inherent in the genetic constitution of glycophytes as well as halophytes and does not apparently arise due to any stable genetic alterations (Bressan et al. 1985; Hasegawa et al. 1986; Kononowicz et al. 1990a; Casas et al. 1991). However, the limits, with respect to the period required for adjustment and the severity of the stress (i.e. salt concentration) that can be accommodated differ amongst genotypes. Similar observations of the salt adjustment capacities of plants have also been made, although it is yet unclear if any stable alterations in genetic expression occur (Amzallag et al. 1990; Bressan and Hasegawa unpubl.).

After adjustment to the saline environment, cells are capable of withstanding greater salt concentrations, i.e. exhibit increased tolerance (Binzel et al. 1985; Hasegawa et al. 1986). This may be due in large part to the "adjusted state" of the cells that predisposes them to withstand greater absolute stress without it being lethal. However, the cells do not seem capable of coping with any greater increase in salt concentration above that to which they are adjusted than they could before adjustment, i.e. the elastic limits of the capacity for physiological adjustment to salinity seems to be tightly controlled (Binzel et al. 1985). Salt tolerance attributable to physiological adjustment is not stable through mitosis after relaxation of the stress and is not transmissible through regeneration (Bressan et al. 1985; Kononowicz et al. 1990a).

Selection in vitro also results in the isolation of cells that express enhanced salt tolerance after the stress is eliminated and the phenotype is stable through regeneration (Mathur et al. 1980; Nabors et al. 1980, 1982; Tyagi et al. 1981; Yano et al. 1982; Bressan et al. 1985, 1987; Nabors and Dykes 1985; Bhaskaran et al. 1986; Pua and Thorpe 1986; McCoy 1987b; Bouharmont and Dekeyser 1989; Hanning and Nabors 1989; Narayanan and Sree Rangasamy 1989; Vajrabhaya et al. 1989; Freytag et al. 1990; Jain et al. 1990, 1991; Waskom et al. 1990; Watad et al. 1991b). Difficulties in isolating cells that possess a genetic capacity to tolerate high salt concentrations from those that are exhibiting an inherent capacity for adjustment into saline environments have been circumvented either by isolating cell clones or through establishment of selection protocols capable of identifying cells with a genetically based predisposition for salt tolerance (Dix and Street 1975; Nabors 1983, 1990; Bressan et al. 1985; van Swaaji et al. 1986; Ochatt and Power 1988; Tal 1990). Salt tolerance of cells adapted to 428 mM (Binzel et al. 1985) and 500 mM (Watad et al. 1983) NaCl is retained after several years of growth in medium without NaCl (Bressan et al. 1985; Watad et al. 1985, 1991b). With tobacco, there seems to be an absolute NaCl concentration that supersedes the inherent genetic capacity for salt acclimation by unadapted cells (Bressan et al. 1985).

Heritable variation for salt tolerance in tobacco cells is the result of population enrichment for cells exhibiting the tolerance phenotype (Bressan et al. 1985). It is not certain to what extent the culture environment, including NaCl, directs any stable genomic alterations (Skokut and Filner 1980; Yamaya and Filner 1981; Handa et al. 1983; Meins 1983; Donn et al. 1984; Cullis and Cleary 1985; Lee and Phillips 1988; Brettell and Dennis 1991) that would enhance salt tolerance. The establishment of the stable NaCl-tolerant phenotype is accompanied by a stable, albeit polyploid, chromosome complement (Kononowicz et al. 1990a, b) and the capacity for increased expression of genes whose products are presumed to be of adaptive benefit to the cells (LaRosa et al. 1989; Perez-Prat et al. 1990; Narasimhan et al. 1991).

Since the report by Nabors et al. (1980), that first indicated somatic cell selection techniques could result in the regeneration of tobacco plants with enhanced salt tolerance and that this phenotype is heritable, considerable effort has gone into applying this technology to other genotypes. The literature has been recently summarised by Tal (1990), although the author is justifiably guarded about whether or not a conclusion can be reached regarding the utility of somaclonal variation for obtaining salt-tolerant plants, particularly from an agricultural perspective. Regardless, it seems clear that plants can be regenerated from cultures growing in vitro that exhibit greater salt tolerance than the original genotypes. Somaclonal variation for salt tolerance has been determined to be inherited in the progeny of tobacco (Nabors et al. 1980; Bressan et al. 1987), sorghum (Bhaskaran et al. 1986; Waskom et al. 1990), flax (McHughen 1987), rice (Vajrabhaya et al. 1989; Narayanan and Sree Rangasamy 1989), *Brassica juncea* (Jain et al. 1990), alfalfa (Winicov 1991), and sugar beet (Freytag et al. 1990) including selections that exhibit enhanced salt tolerance in the R_4 generation. These results seem to indicate that intrinsically cellular mechanisms that are selected for in vitro will confer an enhanced salt tolerance phenotype to regenerated plants.

Many questions about the potential agricultural utility of exploiting somaclonal variation to increase salt tolerance have been recently discussed in the context of our current understanding (Tal 1990). We would like to address this point from the perspective that a principal difficulty in all traditional or nontraditional endeavours directed towards crop improvement for salt tolerance is our fundamental lack of understanding about the interrelationship between osmotic stress tolerance and crop productivity. This makes the choice of selection criteria to obtain salt-tolerant crop genotypes purely intuitive. As a consequence, the effects of such selection strategies on other important phenotypes that affect yield cannot be predicted. Certainly, reduced cell expansion after adjustment to osmotic stress is a universal response of mesophytes and glycophytes (Cutler et al. 1980; Matsuda and Riazi 1981; Meyer and Boyer 1981; Bressan et al. 1982; Michelena and Boyer 1982; Binzel et al. 1985; Termaat et al. 1985; Munns and Termaat 1986; Hsiao and Jing 1987). The response of halophytes would seem to be dissimilar (Waisel 1972; Flowers et al. 1977, 1986; Glenn and O'Leary 1984) but many of these plants cannot be viewed typically as having high capacity for biomass production.

As suggested, there is substantial evidence that indicates salt tolerance and reduced cell expansion are inextricably linked. This may provide some explanation of why many somatic cell selection attempts have yielded regenerants that by some criteria can be classified as salt-tolerant, although the recovered plants exhibit such reduced cell expansion that they are agriculturally undesirable (Nabors et al. 1980; Bressan et al. 1987; McCoy 1987a, b). The impact of reduced cell expansion in young vegetative plants on crop productivity based on the interrelationships of leaf canopy area, photosynthetic assimilation and biomass accumulation can be substantial (Bradford and Hsiao 1982; Kriedemann 1986; Hsiao and Jing 1987).

This does not mean that from an agricultural perspective the potential for using somatic cell selection to obtain salt-tolerant genotypes is entirely pessimistic. There are examples where cell selection techniques have exploited somaclonal variation to obtain genotypes that exhibit yield stability in saline environments (Bhaskaran et al. 1986; McHughen 1987; Waskom et al. 1990). Yield stability of the salt-tolerant genotypes was apparently coupled to increased overall vigour (Bhaskaran et al. 1986; McHughen 1987). Evidence that somaclonal variation for increased drought and perhaps salt stress tolerance can be selected in vitro and the resultant plants and subsequent progeny retain an agronomically acceptable phenotype (Waskom et al. 1990) implies that strategies can be established to circumvent the anomalies that will arise as a result of the high genetic instability in somatic cells in vitro.

Until now, efforts to exploit the potential of somaclonal variation to improve salt tolerance in plants have been restricted to cell selection approaches that are based primarily on empiricism (Nabors et al. 1980; Bhaskaran et al. 1986; Bressan et al. 1987; McHughen 1987; Narayanan and Sree Rangasamy 1989; Vajrabhaya et al. 1989; Freytag et al. 1990; Jain et al. 1990; Tal 1990; Waskom et al. 1990). It is not possible to ascertain exactly what determinants of salt tolerance have been selected and what is the interrelationship, if it exists, between these determinants and reduced agricultural productivity. As indicated, some research results provide reason for optimism (McHughen 1987; Bhaskaran et al. 1986; Waskom et al. 1990). However, in order to realise the full potential of somaclonal variation to improve salt tolerance of crops, it is necessary to devise selection protocols that delineate between the determinants of salt tolerance and those of other phenotypes that detract from the agricultural utility of a genotype.

As suggested, it is possible that some salt tolerance determinants are coupled to characters that are not considered to be agriculturally useful and this linkage is a necessary component of adaptation, albeit not possible to explain based on current knowledge. If an intrinsic coupling between salt tolerance and reduced cell expansion exists, it would have profound detrimental impact on agricultural productivity, particularly if viewed in the context of yields that occur in the absence of or under marginal stress. However, one has quite a different perspective in extreme stress environments, where a substantially reduced yield is certainly more desirable than the typical alternative of no productivity that would be applicable to most crop plants.

Regardless of the extent of coupling between salt tolerance and crop productivity, it is certain that some degree of tolerance can be gained without substantial reduction in yield (Blum 1988). It has been proposed that determinants of salt tolerance may be cryptic in genotypes that are unable to survive or do poorly in response to salt stress based on certain selection criteria but possess agriculturally useful phenotypes, including high harvest indices (Yeo and Flowers 1986; Yeo et al. 1990). This implies no or a low coupling between certain salt tolerance determinants and reduced cell expansion. Identification of precise selection criteria should provide a means to utilise somatic cell selection for pyramiding salt tolerance determinants into agriculturally useful genotypes in a manner that either breaks the linkage with reduced cell expansion or exploits, to the maximum extent, the elastic limits of this linkage.

5.2.2 Parasexual Hybridisation

The use of parasexual hybridisation, mediated through protoplast fusion techniques, for improvement of salt tolerance in plants presently is based primarily on theoretical potential (Finch et al. 1990). Conceptually, cell fusion is a means to hybridise two species that are in tertiary gene pools or are taxonomically unrelated, i.e. cannot be hybridised by alternative plant genetic techniques (Gleba and Sytnik 1984; Glimelius 1988). One substantial limitation of this technology has been the multiple euploidy and aneuploidy of the hybrids resulting from such fusions (Gleba and Sytnik 1984; Glimelius 1988). These hybrids are products of parental genome combinations that arise from nuclear fusion but without the capacity to eliminate precisely and efficiently particular chromosomes after desired recombination events have occurred (Wijbrandi et al. 1990a, b, c; Sundberg et al. 1991). The polyploid genome is not readily amenable to the appropriate genetic manipulations required for agricultural improvement and distribution of many crops that are propagated by seed. Applicability to some plants may be possible, particularly clonally propagated plants, and strategies have been suggested to utilise specifically products of nuclear genome fusions in breeding approaches (Wenzel et al. 1979; Waara et al. 1991).

It is not clear that the products of protoplast fusions between genetically divergent genotypes undergo sufficient nuclear genome recombination to offer realistic potential for the introgression of desirable unlinked genes independent of unwanted genetic background (Wijbrandi et al. 1990b, c). Regardless of the theoretical potential of cell fusion to preclude post-zygotic barriers, phylogenetic limitations to genomic integration would seemingly restrict this potential. Additionally, the lack of specific salt tolerance selection criteria and very limited ability to identify agronomic phenotypes in vitro would seemingly make the potential for obtaining cells and ultimately plants that contain desirable recombinations in the appropriate genetic background rather problematic. Widespread utility of protoplast fusion techniques to obtain nuclear genome hybridisation for the improvement of salt tolerance in crop plants would appear to require some substantial scientific and technical advances.

Protoplast fusion techniques have been used to produce cybrids through the transfer of the mitochondrial and/or chloroplast genomes of a donor into the nuclear background of a recipient genotype (Gleba and Sytnik 1984; Galun and Aviv 1986; Galun et al. 1988; Eigel et al. 1991). The resultant cytoplasm may contain any genomic array, ranging from homologous recipient to homologous donor complements, including heterologous combinations of recipient and donor organelles (Belliard et al. 1978, 1979; Aviv et al. 1984; Gleba and Sytnik 1984; Galun and Aviv 1986; Galun et al. 1988). Selection pressure will, however, enrich the population for a particular genomic complement (Moll et al. 1990; Perl et al. 1991).

Based on the interactions that occur between products of nuclear and cytoplasmic genomes (Taylor 1989; Breiman and Galun 1990), it might be predicted that there are stringent limits on the amount of genetic material that can be transferred without significant phenotypic modulation. Evidence that cybrids can undergo both plastome (Medgysey et al. 1985) and chondriome recombination (Belliard et al. 1979; Boeshore et al. 1985; Barsby et al. 1987; Morgan and Maliga 1987; Perl et al. 1991) is indicative that such limited transfers are feasible. The utility of cybridization techniques to improve salt tolerance in plants will, of course, be dependent on the significance of chondriome and plastome genes to the phenotype (Ramage 1980; Rietveld et al. 1988) and perhaps the capacity to impose very specific selection for salt tolerance determinants in vitro.

5.2.3 Gene Transformation

The recent advances in gene isolation, vector construction and transformation technologies seem to be clear indicators of the advent of the capacity to utilise recombinant DNA techniques to deliver specific genes to host genotypes for agricultural benefit (Alwen et al. 1990; Binns 1990; Caboche 1990; Christou 1990; Lindsey and Jones 1990; Neuhaus and Spangenberg 1990; Sanford 1990; Corbin and Klee 1991; Oard 1991; Potrykus 1991; Rogers 1991). There is evidence to indicate that plants can be genetically engineered and the desired phenotype obtained (Beachy et al. 1987; Fischhoff et al. 1987; de Greef et al. 1989; Kaniewski et al. 1990; Lee et al. 1990; Mariani et al. 1990; Napoli et al. 1990; van der Krol et al. 1990; Broglie et al. 1991; Oeller et al. 1991). The capacity to transform plants with gene constructs that regulate the expression of the transferred gene in the appropriate tissues and under the correct developmental programme and to target the product properly will be highly feasible in the near future (Bednarek et al. 1990; Gilmartin et al. 1990; Hageman et al. 1990; Koltunow et al. 1990; Schmid et al. 1990; Fritze et al. 1991; Samac and Shah 1991; van der Krol and Chua 1991).

Transformation strategies involving direct DNA uptake, electroporation, microinjection, microprojectile bombardment and other emerging technologies will likely overcome the limitations of the natural vector *Agrobacterium* whose host range does not appear to include graminaceous crop plants (Alwen et al. 1990; Binns 1990; Caboche 1990; Christou 1990; Lindsey and Jones 1990; Neuhaus and Spangenberg 1990; Sanford 1990; Corbin and Klee 1991; Oard 1991; Potrykus 1991; Rogers 1991). Research is also underway to direct the site of DNA insertion into the host genome (Paszkowski et al. 1988; Lee et al. 1990; Gal et al. 1991). This information will facilitate homologous recombination in instances where allelic substitutions with transferred genes are possible. Control of insertion sites in the host genome should theoretically reduce the potential of adverse pleiotropic effects on agricultural phenotypes that may arise from random DNA integration. Such effects could remain cryptic in the genotype until subsequent generations, substantially complicating variety development efforts.

It is now also clear that the products of genes encoded by organellar genomes can be targeted to the appropriate organelle and function if the gene constructions contain the appropriate sequences for importation (Hageman et al. 1990). The agricultural efficacy of genotypes transformed with genes encoded by cytoplasmic genomes should be improved as effective techniques for organellar transformation are developed (Svab et al. 1990). Substitution for the wild-type gene by homologous recombination in many instances will substantially enhance the phenotype of the transgenic plant.

What is clear from the perspective of utilising gene transfer technology to improve salt tolerance in plants is the need for precise molecular determinants. These determinants may include genes whose products are functional components of the physiological and biochemical processes required for adaptation to and growth in saline environments. Genes that have regulatory function in the perception and transduction of signals for osmotic stress adaptation or in the co-ordination of processes required for homeostatic growth after adaptation are also of substantial importance.

A variety of strategies are being employed to identify molecular determinants of osmotic stress tolerance (Cushman et al. 1990; Skriver and Mundy 1990). These include characterisation of genes that are upregulated by salt or other inducers of osmotic stress or by elicitors that are presumed to be in the signal transduction pathway of salt stress responses (e.g. abscisic acid). An alternative approach has been to identify genes that are expressed in salt-tolerant genotypes or in genotypes after salt adaptation. Usually, minimal or no information about the function of the gene product is known. In some instances, comparison of predicted peptides with those available in various gene banks has led to tentative identification of function based on significant sequence homology with proteins that have been biochemically characterised. However, some osmotically regulated genes have been isolated that encode products considered to have substantial metabolic function in adaptation to salt stress, including an isogene of phosphoenolpyruvate carboxylase (Cushman et al. 1989, 1990), betaine aldehyde dehydrogenase (BADH: Weretilnyk and Hanson 1990), Δ^1-pyrroline-5-carboxylate reductase (P5CR: Delauney and Verma 1990) and E_1E_2 type and tonoplast proton (H^+)-ATPase (Hasegawa et al. 1990; Perez-Prat et al. 1990; Narasimhan et al. 1991; Surowy and Boyer 1991) genes. The soybean P5CR gene was isolated by direct complementation of *a pro*C mutation in *Escherichia coli*, indicating potential use for heterologous systems to obtain functional molecular determinants of salt adaptation.

Efforts are also being directed toward establishment of causal relationships between molecular and physiological and biochemical determinants of salt tolerance; characterisation of molecular determinant regulation, including identification of controlling elements and factors and assessment of the contribution of putative molecular determinants to salt tolerance. This last objective is being pursued typically through experiments that involve over-and under-expression of transferred genes. However, beside the technical difficulties that are inherent to these experiments, it may not be possible to arrive at a reasonably conclusive interpretation if the gene examined does not encode a product that limits the salt stress response of the host genotype. Significant information may be provided through experiments to complement phenotypically defined salt-susceptible mutants that are available in heterologous systems; particularly until suitable plant mutants are available.

5.3 Tissue Culture in the Identification and Characterisation of Cellular Determinants of Salt Tolerance

Regardless of whether traditional or non-traditional approaches are used, efforts to improve salt tolerance of plants should be facilitated by a definitive understanding of the physiological, biochemical and molecular genetic mechanisms that are involved in adaptation to and growth in the stress environment. This information should form the basis for identification of precise criteria for salt tolerance selection in both plant breeding programmes as well as in strategies involving the use of somatic cell selection, parasexual hybridisation or genetic transformation techniques. So far, progress towards developing salt-tolerant crops has been rather limited. This is attributable not only to limited understanding of the mechanistic determinants of tolerance but also to a rather vague comprehension of the interrelationship between salt tolerance mechanisms and plant growth.

As cellular-based mechanisms are fundamental to the capacity of plants to adapt to and grow in saline environments, then in vitro cultures must express the appropriate genes and utilise the necessary mechanisms required to cope with the stress. Data obtained from cell and tissue cultures indicate the ubiquity of many cellular responses in vitro and in vivo (Heyser and Nabors 1981a, b; Ben-Hayyim and Kochba 1983; Watad et al. 1983, 1986; Binzel et al. 1985, 1987, 1988; LaRosa et al. 1985, 1987; Ben-Hayyim et al. 1987; Bressan et al. 1990; Schnapp et al. 1990, 1991; Tal 1990; Dracup 1991). Over the last several years, we have utilised cultured cells of a glycophyte, tobacco, to dissect cellular determinants of NaCl tolerance (Hasegawa et al. 1986, 1990; Bressan et al. 1987, 1990; Binzel et al. 1989; LaRosa et al. 1989; Rhodes et al. 1989; Singh et al. 1989; Kononowicz et al. 1990a, b; Reuveni et al. 1990; Narasimhan et al. 1991; Schnapp et al. 1991; Watad et al. 1991a, b). A glycophyte was chosen since virtually all crop plants are glycophytes and the assumption that at least some responses to salt are uniquely different from those of a halophyte.

Cells in vitro utilise mechanisms for osmotic adjustment in a saline environment analogous to those used by cells in the actively growing tissues of plants responding to salinity (Flowers et al. 1977, 1986; Epstein 1980; Greenway and Munns 1980; Heyser and Nabors 1981a, b; Wyn Jones 1981; Poljakoff-Mayber 1982; Watad et al. 1983; Wyn Jones and Gorham 1983; Binzel et al. 1985, 1987, 1988, 1989; Hasegawa et al. 1986; Dracup 1991). Cell expansion is perturbed, i.e. reduced, as it is in tissues of actively growing cells of glycophytes during and after adaptation to NaCl (Binzel et al. 1985, 1989; Termaat et al. 1985; Munns and Termaat 1986; Bressan et al. 1987, 1990). This information indicates that definitive inference can be made about the relationship between osmotic adjustment and, in the future, the mechanisms involved in this process, and cell expansion in glycophytes. The remaining portion of this chapter will continue this discussion concerning the interrelationship between osmotic adjustment and reduced growth relative to salt adaptation, based in part on information obtained from research on cultured cells.

5.3.1 Osmotic Adjustment Is a Fundamental Cellular Determinant of Salt Tolerance

Adaptation to salinity involves mechanisms that must alleviate the dual detrimental effects of water deficit and ion toxicity (Flowers et al. 1977, 1986; Greenway and Munns 1980; Polijakoff-Mayber 1982; Wyn Jones and Gorham 1983; Hasegawa et al. 1986). At the level of the cell, this requires co-ordination of mechanisms that mediate osmotic adjustment with those that regulate cytosolic accumulation of the ions effecting the stress. It is now viewed that in cells of higher plants, osmotic adjustment in saline environments is typically effected by the partitioning of ions into the vacuole and extra-cellularly with organic solutes accounting principally for the osmotic potential of the cytoplasm (Flowers et al. 1977, 1986; Greenway and Munns 1980; Wyn Jones 1981; Storey et al. 1983; Binzel et al. 1988, 1989). Ion concentrations in the cytosol are reduced to levels that minimally attenuate transport and metabolic processes.

5.3.2 Osmotic Adjustment Mediated by Ion Accumulation

Osmotic adjustment by higher plant cells, in response to saline environments, invariably involves utilisation of the principal ions in the environment, typically Na^+ and Cl^-, as osmolytes (Flowers et al. 1977, 1986; Yeo 1981; Storey et al. 1983; Binzel et al. 1988). Vacuolar ion accumulation was postulated in these circumstances, since in vitro assays did not indicate that cytosolic enzymes of salt-tolerant plants had greater capacities for catalytic activity in solutions of high ionic strength than those of intolerant plants (Flowers et al. 1977, 1986). Indirect evidence of ion compartmentation was provided through the interpretation of steady-state efflux kinetic analyses of isotopic exchange washout isotherms (Yeo 1981; Cheeseman 1986). Compartmentation was later confirmed by direct measurements of cytosolic and vacuolar ion concentrations using X-ray microanalysis (Storey et al. 1983; Hajibagheri et al. 1987; Binzel et al. 1988) and by determination of ion concentrations in isolated organelles (Schroppel-Meier and Kaiser 1988).

Determinations of Na^+ and Cl^- concentrations in the cytosol and vacuole, albeit somewhat limited in number, are indicative that extremely steep concentration gradients (greater than ten fold) do not exist between the two compartments or between the cytosol and the apoplast. Perhaps the thermodynamic costs of large concentration gradients would severely detract from the energy budget of cells to the extent that growth and morphogenesis in the saline environment would not be feasible or would substantially limit the capacity of the plant to compete effectively in the environment. However, the measured cytosolic concentrations of Na^+ and Cl^- seem sufficient to limit metabolic processes, if it is possible to extrapolate from results obtained with in vitro assays of enzyme activities (Shomer-Ilan and Waisel 1986; Csonka 1989; Manetas 1990; Schwab and Gaff 1990; Shomer-Ilan et al. 1991). This information may indicate the likelihood of the osmoprotectant function in the cytosol of certain organic solutes.

Many salt-tolerant plants exhibit greater K^+/Na^+ selectivity than intolerant plants, as manifested by higher relative leaf K^+/Na^+ ratios, and this capacity has been implicated as a significant salt tolerance adaptation (Shah et al. 1987; Schachtman et al. 1989; Gorham et al. 1990; Wolf et al. 1990, 1991). Greater capacity for K^+/Na^+ selectivity is genetically based, although current information indicates that there may be numerous genetic determinants involved (Shah et al. 1987; Schachtman et al. 1989; Gorham et al. 1990). Physiologically, present evidence does not unequivocally implicate mechanisms that regulate either ion uptake into the xylem or importation into the leaf symplast as the primary effectors of K^+/Na^+ selectivity (Jeschke 1984; Gorham et al. 1990; Wolf et al. 1990, 1991). There is indication that numerous mechanisms may contribute to K^+/Na^+ selectivity and the involvement of these mechanisms may vary with the genotype. The capacity for high K^+/Na^+ selectivity in plants, particularly in the presence of high external Na^+, is strongly facilitated by the external Ca^{2+} concentration (Läuchli and Epstein 1970; Rains 1972; Cramer et al. 1985; Läuchli 1990).

Vacuolar Na^+ and Cl^- compartmentation is apparently tightly co-ordinated with the capacity of the plasma membrane to regulate net uptake of these ions. The $t_{1/2}$ for steady-state exchange of Na^+ and Cl^- across the tonoplast is orders of magnitude longer than the plasma membrane (Pallaghy and Scott 1969; Walker and Pitman 1976; Binzel et al. 1988). In all probability, mechanisms that restrict net uptake of these ions across the plasma membrane and enhance net uptake across the tonoplast must function in a tightly regulated manner to achieve vacuolar compartmentation without toxic accumulation of ions in the cytosol (Reinhold et al. 1989; Hasegawa et al. 1990). This requirement would be substantial during the period immediately after initial exposure to salt.

Mechanisms that restrict intracellular Na^+ uptake contribute to higher K^+/Na^+ selectively, although it is not definitively known if this is a result of reduced influx or increased efflux (or both) across the plasma membrane (Jeschke 1984). The capacity to restrict net Na^+ uptake is modulated by Ca^{2+} (Läuchli 1990). Na^+ uptake into NaCl tolerant tobacco cells is substantially less than into intolerant cells (Binzel et al. 1989; Hasegawa et al. 1990). It is uncertain to what extent K^+ flux differences (i.e. enhanced influx or decreased efflux) influence K^+/Na^+ selectivity, however, enhanced K^+ uptake capacity of NaCl tolerant tobacco cells appeared to be attributable to increased K^+ influx (Watad et al. 1991a).

Establishment and maintenance of steady-state cytosolic and vacuolar Na^+, K^+, Ca^{2+} and Cl^- concentrations are the composite result of the action of pumps, carriers and channels that regulate ion transport across the plasma membrane, tonoplast and organellar and vesicular membranes (Poole 1988; Hedrich and Schroeder 1989; Tester 1990). Co-ordination of the activities of transporters is undoubtedly a requisite for satisfactory intracellular ionic homeostasis in saline environments. Although far from clearly established, some of the interactions can be inferred from current understanding about the direct and indirect coupling of the various transporters.

With the exception of plasma membrane, endoplasmic reticulum and perhaps tonoplast Ca^{2+}-ATPases (Giannini et al. 1987; Robinson et al. 1988; DuPont et al. 1990; Evans et al. 1991) that are presumed to transport Ca^{2+} directly to the apoplast, into the endoplasmic reticulum and into the vacuole, respectively, only H^+ pumps have been identified in plant cells (Spanswick 1981; Sze 1985; Rea and Sanders 1987; Hedrich and Schroeder 1989; Serrano 1989; Sussman and Harper 1989). The presently known plant H^+ pumps are ATPases and a tonoplast-sited pyrophosphatase. These pumps cause the electrogenic formation of H^+ gradients (H^+ electrochemical potential gradients, $\Delta\mu_{H^+}$) by transducing the free energy released from the hydrolysis of high energy phosphate bonds (substrates are primarily ATP and PPi, respectively) to facilitate vectorial H^+ transport. The $\Delta\mu_{H^+}$ provides the free energy for solute transport against electrochemical gradients.

With the exception of H^+ and Ca^{2+}, ion transport in cells of higher plants is currently assumed to be directly mediated through channels or carriers (Poole 1988; Hedrich and Schroeder 1989; Reinhold et al. 1989). The former transport ions passively (down free energy gradients) as a function of the electrical or chemical potential gradient or both. Passive transport of ions often occurs as the result of the electrogenic activity of a H^+ pump, and is thus coupled to an energy-dependent process. Uphill transport, i.e. against electrochemical gradients, couples the movement of ions with the downhill movement of H^+ either in the same (symport) or opposite direction (antiport). Depending on the ion transported and the direction of flux relative to H^+ transport, symports and antiports may or may not be electrogenic. Thus, in plants the activities of H^+ pumps are a requisite for energy-dependent transport through coupling to channels or carriers.

Partial characterisation of Na^+, K^+, Ca^{2+} and Cl^- transport across the plasma membrane provides some indication of how uptake of these ions may be mediated in saline environments (Poole 1988; Hedrich and Schroeder 1989; Reinhold et al. 1989; Tester 1990). Na^+ influx across the plasma membrane occurs down a substantial free energy gradient. A Na^+ specific channel has yet to be identified, thus entry across the plasma membrane is presumed to be mediated through a K^+ channel with some affinity for Na^+ or a cation outward rectifying channel (Schachtman et al. 1991). At low Na^+_{ext} the free energy gradient is primarily the result of the membrane electrical potential. However, at high external Na^+ concentrations a substantial chemical gradient will exist between the apoplast and the cytosol. Converse to influx, efflux of Na^+ couples the inward movement of H^+ down its gradient with uphill outward flux of Na^+

through what is presumed to be a plasma membrane sited Na^+/H^+ antiporter (Braun et al. 1989; Hassidim et al. 1990). Assuming that intracellular import of Na^+ after salt imposition is regulated primarily at the plasma membrane, it is not certain if Na^+ discrimination in tolerant genotypes is attributable to decreased influx or increased efflux or both (Braun et al. 1989; Hassidim et al. 1990; Schachtman et al. 1991). Induction or activation of the plasma membrane sited Na^+/H^+ antiporter by external NaCl has not been detected. Enhanced antiport activity elicited by salt treatment has been attributed to a modification of transport kinetics. Na^+/H^+ antiport activity was detected in vesicles from both a glycophyte and a halophyte (Hassidim et al. 1990).

In typical environments (> 5 mM K^+), K^+ influx across the plasma membrane occurs electrophoretically, driven by an inside negative membrane potential (Bentrup 1990). Energy-dependent K^+ uptake utilises the $\Delta\mu_{H^+}$ established by the H^+-ATPase. Efflux of K^+ through outward rectifying channels occurs down free energy gradients when the plasma membrane is depolarised sufficiently. These channels are postulated to have substantial function in turgor and volume regulation (Hedrich and Schroeder 1989; Bentrup 1990; Tester 1990). Efflux of K^+ via a specific K^+/H^+ antiporter has been hypothesised to represent a short-term control mechanism for the maintenance of pH homoeostasis (Bentrup 1990; Hassidim et al. 1990).

Gating of inward and outward rectifying K^+ channels is voltage-sensitive (Hedrich and Schroeder 1989; Bentrup 1990; Tester 1990); inward rectification occurs when the membrane is hyperpolarised and outward rectification when the membrane is depolarised (i.e. greater or less than about-100 mV). As a consequence, the electrogenic activity of the plasma membrane H^+-ATPase substantially influences K^+ transport through these channels. Reduced cytosolic Ca^{2+} concentrations activate inward rectifying K^+ channels while higher concentrations reduce inward K^+ current (Schroeder and Hagiwara 1989; Tester 1990; Schroeder and Thuleau 1991).

Under physiological conditions, influx of Ca^{2+} across the plasma membrane is mediated by inward rectifying channels that transport the ion down the electrochemical potential gradient resulting from the inside negative membrane potential and substantial difference in Ca^{2+} concentrations between the apoplast and the cytosol (Schroeder and Thuleau 1991). These channels are somewhat non-specific, conducting a number of divalent cations and monovalent cations when they are predominant (Tester 1990; Schroeder and Thuleau 1991). Gating of these channels is voltage-dependent, abscisic acid-activated, stretch-activated and controlled by phosphorylation (Tester 1990; Schroeder and Thuleau 1991).

Ca^{2+} is transported across the plasma membrane to the apoplast directly by a Ca^{2+}-ATPase (Rasi-Caldogno et al. 1987; Robinson et al. 1988; Evans et al. 1991). This ATPase apparently functions in H^+/Ca^{2+} exchange, although it has not been determined if the pump is electrogenic, i.e. the stoichiometry of H^+/Ca^{2+} exchange is unknown, although a H^+/Ca^{2+} stoichiometry of 2 has been suggested based on thermodynamic considerations (Rasi-Caldogno et al.

1987; Evans et al. 1991). The Ca^{2+}-ATPase is regulated by Ca^{2+}-calmodulin directly and not via a phosphorylation mechanism (Evans et al. 1991).

At physiological membrane potentials, influx and efflux of Cl^- across the plasma membrane occurs against and down free energy gradients, respectively. This is most likely the situation even under conditions of high external NaCl concentrations, as it is improbable that the Cl^- chemical potential gradient would be sufficient to drive passive influx unless membrane depolarisation occurs. Cl^- uptake is postulated to be mediated by a $Cl^- - H^+$ symport that has a stoichiometry of 1 Cl^- and 2 H^+ (Poole 1988). Cl^- channels have been identified that are regulated by hyperpolarisation, Ca^{2+} or stretch activation, that are likely involved in efflux of the anion (Tester 1990).

The concentration of Ca^{2+} in the vacuole (three to four orders of magnitude higher concentration than in the cytosol) is the function of a Ca^{2+}/H^+ antiporter (Poole 1988; Evans et al. 1991) that is driven by the H^+ gradient across the tonoplast and a putative Ca^{2+}-ATPase that directly pumps the cation into the vacuole (DuPont et al. 1990). Release of Ca^{2+} from the vacuole is mediated by Ca^{2+} channels that are activated by inositol 1,4,5 triphosphate, i.e. activation is mediated by high cytosolic Ca^{2+} levels (Schroeder and Thuleau 1991).

At high cytoplasmic Ca^{2+} concentrations (greater than 0.3 μM), tonoplast channels are activated at negative voltages (negative inside the vacuole) as well as at slightly positive potentials. The kinetics of activation is relatively slow, hence these are referred to as slow vacuolar (SV) channels (Hedrich and Schroeder 1989). SV channels have a greater permeability for cations than anions, with little discrimination for monovalent cations. SV channels that are highly cation-specific, however, have been described and may be reflective of tissue-associated channel specificity (Hedrich and Schroeder 1989).

Another type of tonoplast channel is activated at decreased cytoplasmic Ca^{2+} (less than 0.3 μM) at either negative or positive potentials. Because the rate of activation of these channels is substantially more rapid than SV channels, these have been termed fast vacuolar (FV) type channels (Hedrich and Schroeder 1989). FV channels are permeable to both cations and anions. It can be envisaged that SV and FV channels mediate the distribution of cations and anions between the cytosol and vacuole down free energy gradients.

Transport of K^+ and Na^+ into the vacuole against the electrochemical potential has been linked to cation/H^+ exchange activity attributable to Na^+/H^+ and K^+/H^+ antiporters (Blumwald and Poole 1985; Garbarino and DuPont 1988; Braun et al. 1989; Fan et al. 1989; Hassidim et al. 1990). These cation/H^+ antiporters have been detected in both glycophytes and halophytes (Hassidim et al. 1990). Induction or activation of the Na^+/H^+ antiporter by salt has been detected (Blumwald and Poole 1987; Garbarino and DuPont 1988), although constitutive antiport activity in non-saline environments may also occur (Hassidim et al. 1990).

The identification and partial characterisation of pumps, channels and carriers that mediate the transport of Na^+, K^+, Ca^{2+} and Cl^- are beginning to provide an understanding of the processes that facilitate ion accumulation and compartmentation during salt adaptation. Information pertaining to how pumps and carriers are activated or induced and how the gating of channels is regulated may provide insight into how cells utilise and integrate the control of these transporters to achieve the appropriate intracellular ion distribution in saline environments.

Regulation of the H^+ pumps is central to the ion transport capacity of cells in response to saline environments. The plasma membrane H^+-ATPase is stimulated by K^+ and requires Mg^{2+} for catalysis (Briskin 1990). This pump is stimulated by auxin, fusicoccin, inositol phospholipids, cytokinins, ABA, light, salinity and is sensitive to changes in turgor (Reinhold et al. 1984; Braun et al. 1986; Watad et al. 1986; Poole 1988; Hedrich and Schroeder 1989; Briskin 1990; Chen and Boss 1991). Recently, it has been determined that increased plasma membrane H^+-ATPase mRNA levels are induced by water stress (Surowy and Boyer 1991) and salinity (Perez-Prat et al. 1990).

The tonoplast H^+-ATPase activity increases substantially as facultative CAM plants respond to NaCl exposure (Struve and Lüttge 1987). Similarly, increased ATPase activity has been detected in tomato roots after exposure to salinity (Sanchez-Aguayo et al. 1991). Adaptation to salinity resulted in a tonoplast ATPase alteration that increased specific H^+ transport and ATP hydrolysis activities (Hasegawa et al. 1990; Reuveni et al. 1990). The mRNA levels of the 70 kDa subunit of the tonoplast ATPase is increased by NaCl (Hasegawa et al. 1990; Narasimhan et al. 1991).

5.3.3 Organic Osmotic Solute Accumulation in the Cytosol

In response to salinity, numerous organic solutes are presumed to accumulate predominantly in the cytosol. Sugars, polyols, amino acids, and quaternary ammonium compounds, e.g. betaines, have been most associated with osmotic adjustment in higher plant cells in response to osmotic stress (Flowers et al. 1977, 1986; Greenway and Munns 1980; Aspinall and Paleg 1981; Wyn Jones 1981; Poljakoff-Mayber et al. 1987; Rhodes 1987). The attributes of these molecules as "compatible" solutes include high solubility in water, relative metabolic inertness and minimal effects on charge balance (Yancey et al. 1982; Somero 1986; Csonka 1989).

It is uncertain whether these "compatible" solutes have osmoprotectant function in maintaining enzyme and membrane stability or only serve to balance the water relations of the cytosol. However, in vitro experiments have determined that polyols, amino acids and quaternary ammonium compounds do facilitate the maintenance of enzyme and membrane stability in solutions of high ionic concentrations (Paleg et al. 1984; Shomer-Ilan and Waisel 1986; Csonka 1989; Manetas 1990; Schwab and Gaff 1990; Shomer-Ilan et al. 1991). There are two contrary mechanisms by which these molecules are proposed to evoke stabilisation of proteins (Csonka 1989). One envisages that the molecules interact with the hydrophobic regions of the protein increasing the hydration shell (Schobert 1977; Schobert and Tschesche 1978). The other is based on evidence that these osmoprotectants do not alter the partial molar volume of the protein (Arakawa and Timasheff 1985). This means that the molecules are excluded from the surface of the protein and do not reduce the water activity directly surrounding it.

In general, indications of the osmoprotectant function of these compounds have been based on experiments conducted at concentrations that are substantially above those assumed to occur typically in the cytosol after osmotic adaptation. However, clear delineation of the osmotic contribution of these solutes in the cytosol cannot be ascertained due to limited information about compartmentation and the activity coefficients of these molecules in this compartment. Recently, it has been determined that, at what are assumed to be cytosolic concentrations, these compounds may have osmoprotectant function, particularly when the enzyme concentrations are at physiological (high) levels (Manetas 1990; Shomer-Ilan et al. 1991).

The regulation of processes that mediate organic solute accumulation in plants in response to a change in the osmotic environment is largely undeciphered. Uptake (Reinhold et al. 1984; Wyse et al. 1986) and synthesis (Kauss 1981, 1983; Bental et al. 1990; Chitlaru and Pick 1991) of organic solutes have been linked to volume reduction and turgor-sensing mechanisms. Intracellular uptake of sugars is turgor-sensitive through a mechanism postulated to involve the plasma membrane H^+-ATPase (Reinhold et al. 1984; Wyse et al. 1986). Turgor reduction mediated by a decrease in the external water potential is followed by a substantial increase in extracellular H^+ extrusion that is attributable primarily to the activity of the plasma membrane H^+-ATPase (Reinhold et al. 1984; Watad et al. 1986, 1991a; Wyse et al. 1986). Extracellular acidification provides the thermodynamic gradient for unidirectional sugar influx mediated by a H^+-sugar symport (Reinhold and Kaplan 1984).

Hyperosomotic shock of *Dunaliella salina* cells results in increased cytosolic orthophosphate (Pi) concentration that has been implicated in the triggering of copious glycerol synthesis for osmoregulation (Bental et al. 1990). The increased cytosolic Pi concentration drives the Pi/triose phosphate translocator to flux Pi into the chloroplast and triose phosphate to the cytosol activating the synthesis of glycerol from starch. Similarly, accumulation of isofloridoside (α-galactosyl-1-> 1-glycerol) in *Poterioochramonas malhamensis* is induced after volume reduction due to increased osmolality of the external solution (Kauss 1981, 1983, 1987). The synthesis of isofloridoside is mediated by the activation of isofloridoside-phosphate synthase through a Ca^{2+}/calmodulin-dependent process involving an enzyme that activates the synthase.

In the sections immediately following, we will consider the regulation of glycinebetaine and proline accumulation in higher plants with special reference to the relative contributions of enhanced synthesis versus reduced catabolism or growth to net accumulation of these nitrogenous solutes. We also discuss the possibility that the accumulation of non-nitrogenous solutes could play important roles in osmotic adjustment when N is limiting, or where maintenance of large pools of nitrogenous solutes could place prohibitive metabolic demands on the N budget of the plant cell. These discussions will not be restricted to tissue culture systems, but where possible we will note the utility of tissue culture systems to explore the regulatory mechanisms and their elastic limits.

5.3.3.1 Glycinebetaine Accumulation

Considerable evidence indicates that glycinebetaine accumulates in the cytoplasm and chloroplasts of species of the family Chenopodiaceae during osmotic stress and acts as a non-toxic or compatible osmolyte (Pollard and Wyn Jones 1979; Wyn Jones 1984; Robinson and Jones 1986; McCue and Hanson 1990). If accumulated primarily in the chloroplast (Robinson and Jones 1986), glycinebetaine could play a particularly important role in stabilising enzymes involved in photosynthesis (Incharoensakdi et al. 1986). It is likely that glycinebetaine plays the same role in other angiosperm families, in which it accumulates in response to stress (Wyn Jones and Storey 1981; Weretilnyk et al. 1989; Hanson et al. 1991), although in these families the subcellular site of glycinebetaine accumulation remains to be determined. In the Chenopodiaceae, glycinebetaine is synthesised in the chloroplast from choline in a two-step oxidation via the intermediate betaine aldehyde. These reactions are catalysed by a ferredoxin-dependent choline monooxygenase and an NAD^+-dependent betaine aldehyde dehydrogenase (Hanson et al. 1985; Lerma et al. 1988; Brouquisse et al. 1989; Weretilnyk and Hanson 1990). The Chenopodiaceae and Poaceae appear to differ with respect to the origin of choline utilised in glycinebetaine synthesis. In the Chenopodiaceae, choline appears to be derived from phosphorylcholine, whereas in the Poaceae, choline appears to be derived primarily from phosphatidylcholine, (Hitz et al. 1981; Hanson and Rhodes 1983). Glycinebetaine appears to be a relatively metabolically inert end product which is not actively catabolised in plants (Hanson and Hitz 1982; McCue and Hanson 1990). Because glycinebetaine is not appreciably catabolised, control of glycinebetaine must reside at the level of synthesis rate and the rate of pool dilution by growth.

Genetic evidence for an osmoregulatory function for glycinebetaine has been presented for barley (Grumet and Hanson 1986). Barley isopopulations differing for glycinebetaine accumulation capacity exhibit significant differences in their capacity to adjust osmotically (Grumet and Hanson 1986). *Zea mays* has a modest capacity for glycinebetaine accumulation in comparison to other members of the Poaceae (Hitz and Hanson 1980). Glycinebetaine is accumulated primarily in young expanding maize leaves (Rhodes et al. 1989), but is not accumulated in maize roots (Rhodes et al. 1989; Voetberg and Sharp 1991). A number of glycinebetaine-deficient genotypes of *Zea mays* have recently been identified (Rhodes and Rich 1988; Brunk et al. 1989; Rhodes et al. 1989; Lerma et al. 1991). Glycinebetaine deficiency in maize is conditioned by recessive alleles of a single locus in 13 glycinebetaine-deficient maize inbreds and selections so far evaluated (Lerma et al. 1991). Glycinebetaine deficiency is associated with a metabolic lesion in the capacity to oxidise choline to betaine aldehyde, but not betaine aldehyde to glycinebetaine (Lerma et al. 1991). This implies that glycinebetaine deficiency may be due to a mutation in the structural gene encoding choline monooxygenase, although a mutation in a regulatory locus cannot yet be precluded (Lerma et al. 1991). Glycinebetaine levels of maize hybrids are proportional to dominant allele dosage (Table 5.1). Thus, hybrids derived from two glycinebetaine-containing inbreds or selections exhibit approximately twice the glycinebetaine level of hybrids derived from glycinebetaine-deficient and glycinebetaine-containing inbreds or selections, regardless of whether the glycinebetaine-deficient parent is used as either male or female (Table 5.1).

Two near-isogenic maize hybrids differing for a single copy of the dominant allele conferring glycinebetaine accumulation have been developed (Rhodes et al. 1989). The glycinebetaine-accumulating hybrid (1720Bb) exhibits significantly less shoot growth inhibition in response to salinity stress in comparison to the glycinebetaine-deficient near isogenic hybrid (1720bb; Table 5.2). Moreover, the betaine-containing hybrid (1720Bb) exhibits less severe symptoms of leaf firing in response to salinity stress (unpublished observations of D. Rhodes and P. J. Rich). These results suggest that even a modest capacity for glycinebetaine accumulation may play an adaptive role in maintaining shoot growth in saline environments in maize.

Glycinebetaine accumulation in leaf tissue of salinized maize plants occurs in proportion to the decline in leaf relative water content associated with salinity

Table 5.1. Glycinebetaine levels of maize hybrids (W.-J. Yang, Y. Samaras, G. C. Ju, P. J. Rich, A. D. Hanson and D. Rhodes, unpubl. results)

Hybrid type[a] (female×male)	Mean betaine level ($\pm$ SD)[b] μmol/g fw	No of hybrids
Bet − × Bet −	0.055($\pm$ 0.042)	55
Bet + × Bet −	2.743($\pm$ 1.303)	37
Bet − × Bet +	3.047($\pm$ 1.422)	31
Bet + × Bet +	5.791($\pm$ 2.225)	20

[a] Glycinebetaine-deficient (Bet^-) inbreds and selections employed as parents included 13 allelic betaine-deficient types; A188, A641, A656, B37, H84, H95, N6, Oh43, Red South American, 2708, 729–13F_3, 86M–9S_2 and 86W–9S_2 (Lerma et al. 1991). Glycinebetaine-positive (Bet^+) inbreds and selections included; A554, A632, B14A, B68, B73, H49, H100, Mo17, MS71, WF9, W64A and 729–04F_3 (Brunk et al. 1989; Lerma et al. 1991).

[b] Each hybrid was evaluated for glycinebetaine level by the periodide assay in triplicate, sampling from young expanding leaves of field grown plants 9 weeks after planting, essentially as described by Lerma et al. (1991).

Table 5.2. Shoot fresh weights of salinized plants of betaine-containing (1720Bb) and betaine-deficient (1720bb) near isogenic sweet corn hybrids, expressed as a percentage of shoot fresh weights of non-salinized plants. (G. C. Ju, P. J. Rich, A. D. Hanson and D. Rhodes unpubl. results)

Hybrid[a]	Shoot fresh weight of salinized plants as % of shoot fresh weight of non-salinized plants[b]			
	Expt1	Expt2	Expt3	Mean(SE)
1720bb	30.5	34.5	34.8	33.3(1.13)
1720Bb	36.1	47.6	38.5	40.7(2.86)*

* Denotes significant difference between hybrids at the $P = 0.05$ level.

[a] See Rhodes et al. (1989) for origin of these hybrids and details of their glycinebetaine levels under non-salinized and salinized conditions.

[b] In each of three independent experiments, populations were grown for 3 weeks in the absence of salinization, half the population of each genotype was then maintained under well-irrigated conditions in the absence of salinization, and half was irrigated with 50 mM NaCl for 1 week, followed by 100 mM NaCl for 1 week, and 150 mM NaCl for 1 week. Plants were harvested for shoot fresh weight determination after 6 weeks.

(Rhodes et al. 1989). This accumulation cannot be solely accounted for by decreased growth, and must involve increased synthesis, since net glycinebetaine content per plant is increased 3.5-fold by salinity (150 mM NaCl) in betaine-containing genotypes (but not betaine-deficient genotypes; Rhodes et al. 1989). Weretilnyk and Hanson (1989) note that betaine aldehyde dehydrogenase (BADH) activity in spinach leaves is increased up to three fold in salinized plants, and is correlated linearly with solute potential; a relationship which cannot readily be explained by models that invoke turgor sensing. Salinity stress not only increases BADH enzyme activity and BADH protein level, but also BADH mRNA level, suggesting transcriptional control of the BADH gene in relation to osmotic stress (Weretilnyk and Hanson 1990). Choline monooxygenase level also appears to be increased by salinity stress in spinach, suggesting co-ordinate regulation of the pathway (Brouquisse et al. 1989).

Thus far, there have been relatively few quantitative studies of glycinebetaine biosynthesis in relation to salinity in tissue cultured plant cells. Suspension cultured cell systems offer considerable advantages over whole-plant systems in assessing the relative contributions of reduced growth rate and increased synthesis rate to the net expansion of the pools of specific solutes. Unadapted (S-0) and 342 mM NaCl adapted (S-20) cell suspensions of the halophyte *Atriplex*

Table 5.3. Glycinebetaine levels and fresh weight doubling times of *Atriplex nummularia* cells adapted to 0 (S-0) and 20 (S-20)g/l NaCl. (Y. Samaras, W.-J. Yang and D. Rhodes, unpubl. results)

Cell culture	Medium NaCl (mM)	Glycinebetaine[a] level (nmol/g fw)	Fresh weight[b] doubling time (h)
S-0	0	1851	44.35
S-20	342	11340	95.33

[a] Glycinebetaine was determined by stable isotope dilution fast atom bombardment mass spectrometry of *n*-propyl betaine esters, essentially as described by Rhodes et al. (1987, 1989).
[b] Fresh weight doubling times are maximum doubling times computed from data presented in Fig. 3A of Casas et al. (1991).

nummularia have recently been derived (Casas et al. 1991). Analyses of glycinebetaine pools of these cultures have indicated that S-20 *Atriplex* cells exhibit a 6.13-fold greater glycinebetaine pool than S-0 *Atriplex* cells (Table 5.3). Because S-20 cells exhibit a 2.15-fold greater fresh weight doubling time than S-0 cells (Table 5.3; data derived from Casas et al. 1991), it follows that the glycinebetaine synthesis rate must be increased approximately 2.85-fold in S-20 cells relative to S-0 cells, in order to maintain a 6.13-fold greater glycinebetaine pool size. Such synthesis rate increases associated with salinity stress are consistent with estimates of effects of salinisation on the induction of choline monooxygenase and betaine aldehyde dehydrogenase enzyme levels discussed above. Observed rates of oxidation of d_9-choline chloride to d_9-glycinebetaine for S-0 (Fig. 5.1A) and S-20 (Fig. 5.1B) *Atriplex* cells only slightly exceeded the expected rates of oxidation based on the assumption that glycinebetaine is synthesised at the rate required simply to maintain the observed pool sizes with growth at the doubling times specified in Table 5.3. Thus, there is little evidence to indicate that adaptation to salinity in *Atriplex* cells leads to more than a two to three-fold increase in the capacity to oxidise choline to glycinebetaine. This illustrates that reduced growth rate and increased synthesis rate may contribute approximately equally to glycinebetaine pool expansion in response to adaptation to salinity in cell suspension cultures of an halophytic member of the Chenopodiaceae.

5.3.3.2 Proline Accumulation

Proline accumulation in response to salinity and/or water stress has been extensively documented in higher plants (Stewart and Larher 1980; Stewart 1981; Hanson and Hitz 1982; Rhodes 1987). Like glycinebetaine, proline may afford osmoprotection (Csonka 1989), and may be selectively accumulated in cytosolic compartments in response to osmotic stress, as demonstrated by analysis of proline contents of vacuolar and extravacuolar fractions of protoplasts isolated from leaves of control and water-stressed *Nicotiana rustica* plants

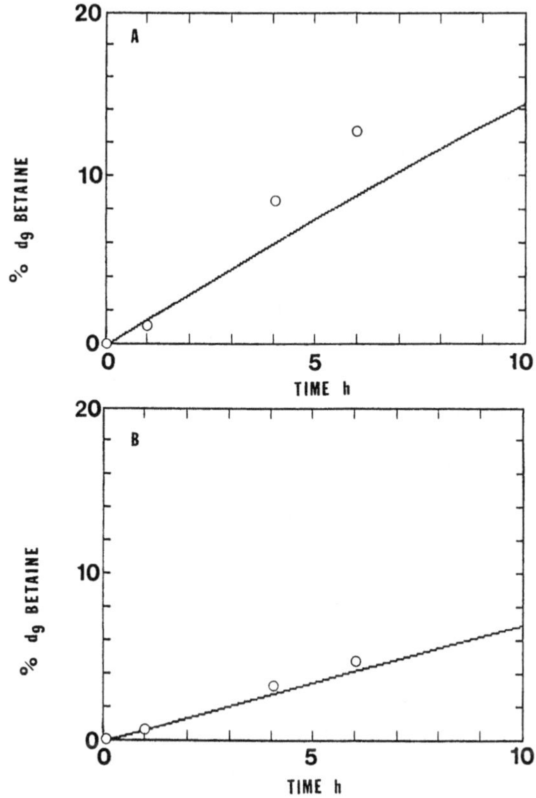

Fig. 5.1A, B. Metabolism of 0.5 mM d_9-choline chloride to d_9-glycinebetaine in S-0 (**A**) and S-20 (**B**) *Atriplex nummularia* cells. Observed data (*open circles*) is expressed as the percentage of the total pool of glycinebetaine accounted for by d_9-glycinebetaine (determined by fast atom bombardment mass spectrometry of *n*-butyl betaine esters, essentially as described by Rhodes and Rich, 1988). *Solid lines* are computer-simulated labelling patterns expected if the glycine betaine pools specified in Table 5.3 were replaced by d_9-glycinebetaine, beginning at $t = 0$ h, at the rates simply required to maintain the free pools with growth at the doubling times specified in Table 5.3. (Y. Samaras, W.-J. Yang and D. Rhodes unpubl. results)

(Pahlich et al. 1983). The prevailing evidence is that osmotic stress primarily leads to an increase in proline synthesis rate from glutamic acid, that this increase in synthesis rate may be due to increased conversion of glutamate to Δ^1-pyrroline-5-carboxylate (perhaps mediated by loss of feedback control of proline synthesis by proline), and that decreased rates of proline oxidation and utilisation in protein synthesis, coupled with increased rates of proteolysis may also contribute to net proline accumulation (reviewed by Stewart 1981).

In maize apical root regions, proline may be the principal organic osmotic solute. Proline concentration increased 10-to 40-fold in apical regions of the root tip at a ψ_w of -1.6 MPa, accounting for up to half the decrease in osmotic potential (ψ_s) (Voetberg and Sharp 1991). The ψ_s contribution of proline in the apical 2 mm, where elongation was fully maintained, was equal to the magnitude of the value of turgor measured with a pressure probe (Spollen and Sharp 1991; Voetberg and Sharp 1991). The accumulation of proline is probably essential in the maintenance of cell expansion in this region. The large increase in proline concentration in the apical region at low ψ_w resulted primarily from increased rates of net proline deposition, rather than decreased volume expansion (Voetberg and Sharp 1991), but the precise biochemical basis of the increased deposition rate (increased synthesis, decreased catabolism, and/or increased uptake) has not yet been determined in this system.

Suspension cell cultures adapted to various osmotic stress regimes provide unique opportunities to assess precisely the relative contributions of increased synthesis, decreased growth rate, decreased protein synthesis, increased proteolysis and decreased proline catabolism, to net proline accumulation. Quantitative flux analysis of nitrogen metabolism in unadapted (S-0) and 428 mM NaCl salt-adapted (S-25) suspension cell cultures of tobacco has resulted in the rate calculations summarised in Table 5.4 (derived from Rhodes and Handa 1989). More than an 80-fold increase in pool size of proline is achieved by an 11-fold increase in proline synthesis rate, a 4-fold reduction in growth rate (i.e. a 4-fold increase in fresh weight doubling time) and a relatively small reduction in the rate of utilisation of proline in protein synthesis. Surprisingly, the rate of catabolism of proline appears to be increased rather than decreased in response to adaptation to salinity stress (Rhodes and Handa 1989).

It is important to note that the total pool size of proline will be determined by the difference between synthesis and utilisation rates. In a steady state, where the proline pool size remains constant with growth, the rate of synthesis of proline (r) plus the rate of proline production from protein turnover (pt) will equal the rate of catabolism of proline (c) plus the rate of utilisation of proline in

Table 5.4. Fresh weight doubling times, pool sizes of proline and estimated rates of synthesis and utilization of proline in tobacco cells adapted to 0 g/l NaCl (S-0) and 25 g/l NaCl (S-25). (After Rhodes and Handa 1989)

	Cell culture	
	S-0	S-25
Fresh weight doubling time (h)	47	196
Proline pool size (nmol/g fw)	1038	83589
Rate of synthesis of proline (r) (nmol/h. g fw)	90	1000
Rate of release of proline from protein via protein turnover (pt) (nmol/h. g fw)	17	3
Rate of utilization of proline in protein synthesis (p + pt) (nmol/h. g fw)	51	12
Rate of catabolism of proline (c) (nmol/h. g fw)	41	695
Proline expansion flux (ef) (nmol/h. g fw)	15	296

protein synthesis (p + pt) (where p is the theoretical rate of synthesis of proline required to maintain a constant pool of proline amino acid residues in protein with growth), and a residual term, the expansion flux (ef) (Kacser 1983; Rhodes and Handa 1989). In a steady state, the expansion flux (ef) = (r + pt) − c − p − pt = r − c − p, which is in turn equal to the pool size of proline multiplied by the natural logarithm of 2(ln 2 = 0.693) divided by the doubling time, thus:

$$\text{ef(nmol/h.g fw)} = \frac{\text{free pool size of proline (nmol/g fw)} \times \ln 2}{\text{doubling time (h)}}.$$

The expansion flux is the rate required to maintain the free pool with steady-state growth, but in non-steady-state situations (where either growth rate or metabolic rates are perturbed) this flux will determine the rate at which the pool will expand or deplete. An expansion flux of only 15.3 nmol/h/g fw is required to maintain a proline pool size of 1038 nmol/g fw in S-0 tobacco cells with a fresh weight doubling time of 47 h, whereas an expansion flux of 296 nmol/h/g fw is required to maintain a proline pool size of 83589 nmol/g fw in S-25 cells with a fresh weight doubling time of 196 h (Table 5.4). Net rates of N assimilation of S-0 and S-25 tobacco cells have been determined to be 7300 and 3180 nmol/h/g fw, respectively (Rhodes and Handa 1989). Thus, proline pool maintenance consumes approximately 0.2% of the N budget of S-0 cells, but over 9% of the N budget of S-25 cells.

These types of calculations raise important questions concerning the adaptive significance of growth inhibition during and after adaptation to osmotic stress, alluded to in earlier sections. For example, it can be readily seen that if S-25 cells were to establish a growth rate (doubling time) equivalent to S-0 cells, then in order to maintain a proline pool size of 83589 nmol/g fw at a doubling time of 47 h, would require a proline expansion flux of 1232 nmol/h/g fw, which would now consume 39% of the N assimilated by these cells (if N assimilation rate were maintained at 3180 nmol/h/g fw). It can be argued that the metabolic costs of maintaining such a large proline pool with a doubling time equal to that of unadapted cells could be prohibitive to the synthesis of other essential N solutes ultimately derived from glutamic acid. It is likely that if the intrinsic coupling between reduced growth and adaptation to salinity stress in tobacco cells is to be broken, several further adjustments in metabolism must be effected. These adjustments in metabolism can be envisaged as follows.

1. If osmotic adjustment with organic solutes is to be achieved primarily with a nitrogenous metabolite such as proline, the expansion flux for this solute must be increased a further four to five fold; this would require either significant reduction in proline catabolism rate and/or further increases in proline synthesis rate.
2. Coupled with these changes in proline synthesis and/or catabolism rate must come increased N assimilation rates required to sustain increased proline expansion fluxes, implying increased rates of NO_3^- or NH_4^+ uptake, and/or increased flux via the nitrate reductase/nitrite reductase assimilatory pathway

and/or the glutamine synthetase/glutamate synthase cycle. Note also that for each mole of proline accumulated, one mole of ATP and two moles of NADPH are consumed (Adams and Frank 1980; Stewart 1981), so that increased growth rates would place proportionately increased demands on ATP and NADPH synthesis required to sustain proline levels.

We propose from such considerations that growth inhibition may be an adaptive mechanism for facilitating osmotic adjustment with nitrogenous organic solutes, such as proline. Slow growth constitutes a mechanism for maintaining elevated pools of organic nitrogenous solutes at relatively low metabolic cost.

Proline biosynthesis may be modulated by ABA (Stewart 1980; Stewart et al. 1986; Pesci 1987, 1989). Pesci and Beffagna (1985) suggest that the ABA-and isobutyric acid-induced changes in proline level in barley leaves might be mediated by changes in the intracellular pH. Thus, both ABA and isobutyric acid cause a marked decrease in intracellular pH and accumulation of proline. Fusicoccin suppresses the ABA-induced acidification, but only partially counteracts the acidifying effect of isobutyric acid. Fusicoccin completely suppresses the ABA-induced increase in proline during short treatment periods, whereas it is only effective in inhibiting the isobutyric acid-induced proline accumulation after long treatment periods (Pesci and Beffagna 1985). ABA-stimulated osmotic adjustment has been observed in tobacco cells, and this may facilitate adaptation to NaCl (LaRosa et al. 1987). The decrease in solute potential caused by ABA treatment in the presence of NaCl is primarily due to increased concentrations of sugars and proline. ABA effects on proline accumulation appear to be largely due to alterations in ψ_s; proline level is a logarithmic function of ψ_s in both tobacco (LaRosa et al. 1987) and tomato (Handa et al. 1986) cells. Simulation studies indicate that for every tenth MPa decrease in ψ_s, the rate of synthesis (r) minus catabolism (c) of proline increases by approximately 4 nmol/h/g fw (Rhodes and Handa 1989). This relationship holds for cell cultures adapted to various levels of salinity stress, as well as for changes in proline level induced by transition from 0 to 10 g/l NaCl in the presence or absence of ABA (Rhodes and Handa 1989). As in the case of glycinebetaine (Weretilnyk and Hanson 1989), such relationships between pool size and rate of synthesis (minus catabolism) of organic solutes and ψ_s are difficult to explain by models which invoke turgor sensing. It is noteworthy that if the linkage between growth inhibition and adaptation to salinity stress is to be effectively broken in tobacco cells, and if proline continues to be utilised as the major organic solute for osmotic adjustment in these cells, then this may demand that fundamental changes occur in the regulation of proline synthesis and/or catabolism as a function of ψ_s.

In tobacco cells adapted to NaCl, increased rates of proline synthesis cannot be attributed to modulation of Δ^1-pyrroline-5-carboxylate reductase enzyme activity (LaRosa et al. 1991), suggesting that control of proline synthesis in relation to osmotic stress must reside at the level Δ^1-pyrroline-5-carboxylate synthesis from glutamate, as also concluded by Stewart (1981). Nevertheless, it should be noted that the levels of mRNA encoding Δ^1-pyrroline-5-carboxylate reductase were higher in leaves and roots of seedlings treated with 400 mM NaCl than untreated plants (Delauney and Verma 1990), suggesting that salinity stress-induced proline accumulation may in part involve regulation of transcription of the gene encoding Δ^1-pyrroline-5-carboxylate reductase.

Major limitations to our understanding of the regulation of proline biosynthesis in higher plants are that the putative first two enzymes of proline biosynthesis, γ-glutamyl kinase and γ-glutamyl phosphate reductase (Adams and Frank 1980), have not yet been unequivocally identified and characterised from any higher plant source (Stewart 1981). However, a recent report of the isolation of a gene from tomato encoding a bi-functional enzyme capable of complementing *Escherichia coli* deletion mutants lacking both γ-glutamyl kinase and γ-glutamyl phosphate reductase (Garcia-Rios et al. 1991), suggest that these limitations may soon be overcome. Proline-accumulating suspension cultured cells adapted to various osmotic stress regimes may provide opportunities to equate precisely changes in rates of synthesis of proline (Rhodes et al. 1986; Rhodes and Handa 1989) to changes of proline biosynthesis gene expression.

5.3.3.3 Accumulation of Other Organic Solutes

In certain members of the Plumbaginaceae, β-alaninebetaine and choline-O-sulphate may supplement or supplant glycinebetaine as a cytoplasmic osmolyte (Hanson et al. 1991). *Limonium* species accumulate either glycinebetaine or β-alaninebetaine, but not both (Hanson et al. 1991). Glycinebetaine and choline-O-sulphate share the same precursor, choline, and thus each potentially interferes with the synthesis of the other. Because choline-O-sulphate may serve to detoxify sulphate, such interpathway competition may be disadvantageous in sulphate-rich habitats (Hanson et al. 1991). The accumulation of β-alaninebetaine rather than glycinebetaine may serve to avoid this interpathway competition. Because β-alaninebetaine can be synthesised in roots, and unlike glycinebetaine does not have an oxygen requirement for synthesis (Brouquisse et al. 1989), it is proposed that β-alaninebetaine accumulation may be advantageous under the severe hypoxia that often prevails in salt marshes (Hanson et al. 1991).

Salt marshes have high rates of sulphide production due to dissimilatory sulphate reduction (Steudler and Peterson 1984). Certain salt marsh species such as *Spartina alterniflora* may possess adaptive mechanisms which permit growth in sediments where dissolved sulphide concentrations exceed those that are toxic to other species (Carlson and Forrest 1982). β-dimethylsulphoniopropionate (DMSP) is accumulated to high levels by *Spartina alterniflora*, and this may represent a mechanism for detoxification of excess sulphide (Dacey et al. 1987). DMSP may represent one of the major sources of dimethylsulphide emitted from salt marshes; DMSP can be readily metabolised to dimethylsulphide and acrylic acid (Dacey et al. 1987).

In many salt marsh communities, nitrogen appears to be the major nutrient limiting growth (Stewart and Rhodes 1978; Stewart et al. 1979; Drake and Gallagher 1984). The accumulation of nitrogenous solutes such as proline and glycinebetaine for osmotic adjustment could, therefore, place high demands on a relatively scarce nutrient resource. It has been suggested that DMSP accumulation by *Spartina anglica* may afford a flexible mechanism of achieving osmotic adjustment under conditions of N deficiency. Thus, it has been proposed that DMSP could possibly substitute for proline and glycinebetaine when N is limiting, and where the levels of nitrogenous-compatible solutes could not be maintained without major changes in protein levels (Stewart et al. 1979). Consistent with this hypothesis, DMSP level increases in proportion to sediment salinity in *Spartina alterniflora*, and tends to decrease with N fertilisation, whereas the concentration of glycinebetaine increases (Dacey et al. 1987). Accumulation of polyols (e.g. glycerol, mannitol or sorbitol) may also provide mechanisms for osmotic adjustment with cytosolic osmolytes whose syntheses are essentially independent of the availability of reduced N. In this context, it may be significant that tobacco cells adapted to NaCl also accumulate sucrose in addition to proline as a major organic osmotic solute (Binzel et al. 1989). Intracellular sucrose and proline concentrations of S-25 tobacco cells are estimated to be 103 and 129 mM, respectively, and together these two solutes account for 77% of the total organic solutes accumulated by tobacco cells. Presumably, sucrose is simply accumulated from the growth medium by tobacco cells, at far less metabolic cost than the cost of de novo synthesis of an equivalent concentration of proline.

It is clear that certain halophytes may have evolved mechanisms for coping with conditions in addition to high concentrations of NaCl which are prevalent in salt marshes (i.e. low N availability, toxic levels of sulphide, anoxia, and high sulphate concentrations). The limitations of tissue culture in mimicking the precise environmental conditions prevailing in natural saline environments should be noted; selection for tolerance to NaCl at the tissue culture level may not necessarily condition tolerance of these other environmental stresses. It should be further emphasised that although accumulation of sucrose from the growth medium may be a highly appropriate adaptive strategy for achieving osmotic adjustment with organic solutes (independently of N metabolism) in suspension cultured cells, such strategies will have obvious limitations in regenerated plants, where accumulation of carbon skeletons from the growth medium is not an available option.

5.4 Conclusion

In this chapter, we have given several explanations of the consistent observation that slower growth rate and adaptation to a saline environment are linked. These explanations, however, all centre on the possibility that some metabolic process that is essential for survival in the saline environment becomes limiting to the cell at normal growth rates. It is therefore a survival advantage to the cell to link closely, presumably through some metabolic co-regulation, reduced growth to the adaptation process. In order to produce rapidly growing salt-adapted cells, it will become necessary to break this linkage. This will require first the identification of changes in those metabolic processes that are actually enhancing the fitness of cells under stress, followed by some genetic and/or environmental alteration that overcomes the limitation on growth. The difficulty of removing any given metabolic limitation will depend on the nature of the limitation, but it is clear that specific metabolic changes in plants that effect fitness during salt stress are yet to be unequivocally identified. Tissue culture technology may be very useful for the definitive identification of such changes.

Acknowledgements. This work was in part supported by grants from the USDA (90-37280-5721 and 91-37100-5873). Journal paper 13727 of the Purdue University Agricultural Experiment Station.

References

Adams E, Frank L (1980) Metabolism of proline and the hydroxyprolines. Annu Rev Biochem 49: 1005–1061

Alwen A, Eller N, Kastler M, Benito-Moreno RM, Heberle-Bors E (1990) Potential of in vitro pollen maturation for gene transfer. Physiol Plant 79: 194–196

Amar L, Reinhold L (1973) Loss of membrane transport ability in leaf cells and release of protein as a result of an osmotic shock. Plant Physiol 51: 620–625

Amzallag GN, Lerner HR, Poljakoff-Mayber (1990) Induction of increased salt tolerance in *Sorghum bicolor* by NaCl pretreatment. J Exp Bot 41: 29–34

Arakawa T, Timasheff SN (1985) The stabilization of proteins by osmolytes. Biophys J 47: 411–414

Aspinall D, Paleg LG (1981) Proline accumulation: physiological aspects. In: Paleg LG, Aspinall D (eds) The physiology and biochemistry of drought resistance in plants. Academic Press, Sydney, pp 205–241

Aviv D, Arzee-Gonen P, Bliechman S, Galun E (1984) Novel alloplasmic Nicotiana plants by donor-recipient protoplast fusion. Cybrids having *N. tabacum* or *N. sylvestris* nuclear genome and either both plastome and chondriomes from alien species. Mol Gen Genet 196: 244–253

Barsby TL, Yarrow SA, Kemble RJ, Grant I (1987) The transfer of cytoplasmic male sterility to winter-type oilseed rape (*Brassica napus* L.) by protoplast fusion. Plant Sci 53: 243–248

Bayliss MW (1980) Chromosomal variation in plant tissues in culture. In: Vasil IK (ed) Perspectives in plant cell and tissue culture. Int Rev Cytol (Suppl) 11A: 113–144

Beachy RN, Abel PP, Nelson RS, Rogers SG, Fraley RT (1987) Transgenic plants that express the coat protein gene of TMV are resistant to infection by TMV. UCLA Symp Mol Cell Biol 48: 205–213

Bednarek SY, Wilkins TA, Dombrowski JE, Raikhel NV (1990) A carboxyl-terminal propeptide is necessary for proper sorting of barley lectin to vacuoles of tobacco. Plant Cell 2: 1145–1155

Belliard G, Pelletier G, Vedel F, Quetier F (1978) Morphological characteristics and chloroplast DNA distribution in different cytoplasmic parasexual hybrids of *Nicotiana tabacum.* Mol Gen Genet 165: 231–237

Belliard G, Vedel F, Pelletier G (1979) Mitochondrial recombination in cytoplasmic hybrids of *Nicotiana tabacum* by protoplast fusion. Nature 281: 401–403
Ben-Hayyim G, Kochba J (1982) Growth characteristics and stability of tolerance of citrus callus cells subjected to NaCl stress. Plant Sci Lett 27: 87–94
Ben-Hayyim G, Kochba J (1983) Aspects of salt tolerance in a NaCl-selected stable cell line of *Citrus sinensis*. Plant Physiol 72: 685–690
Ben-Hayyim G, Kafkafi U, Ganmore-Neumann R (1987) Role of internal potassium in maintaining growth of cultured *Citrus* cells on increasing NaCl and $CaCl_2$ concentrations. Plant Physiol 85: 434–439
Bental M, Pick U, Avron M, Degani H (1990) The role of intracellular orthophosphate in triggering osmoregulation in the alga *Dunaliella salina*. Eur J Biochem 188: 117–122
Bentrup F-W (1990) Potassium ion channels in the plasmalemma. Physiol Plant 79: 705–711
Bhaskaran S, Smith RH (1990) Regeneration in cereal tissue culture: a review. Crop Sci 30: 1328–1336
Bhaskaran S, Smith RH, Schertz KF (1986) Progeny screening of *Sorghum* plants regenerated from sodium chloride-selected callus for salt tolerance. J Plant Physiol 122: 205–210
Binns AN (1990) Agrobacterium-mediated gene delivery and the biology of host range limitations. Physiol Plant 79: 135–139
Binzel ML, Hasegawa PM, Handa AK, Bressan AK (1985) Adaptation of tobacco cells to NaCl. Plant Physiol 70: 118–125
Binzel ML, Hasegawa PM, Rhodes D, Handa S, Handa AK, Bressan RA (1987) Solute accumulation in tobacco cells adapted to NaCl. Plant Physiol 84: 1408–1415
Binzel ML, Hess FD, Bressan RA, Hasegawa PM (1988) Intracellular compartmentation of ions in salt-adapted tobacco cells. Plant Physiol 86: 607–614
Binzel ML, Hess FD, Bressan RA, Hasegawa PM (1989) Mechanisms of adaptation to salinity in cultured glycophyte cells. In: Cherry JH (ed) Biochemical and physiological mechanisms associated with environmental stress tolerance. NATO ASI Series, vol G19. Springer, Berlin Heidelberg New York, pp 139–157
Blum A (1988) Plant breeding for stress environments. CRC, Boca Raton
Blumwald E, Poole RJ (1985) Na^+/H^+ antiport in isolated tonoplast vesicles from storage tissue of *Beta vulgaris*. Plant Physiol 78: 163–167
Blumwald E, Poole RJ (1987) Salt tolerance in suspension cultures of sugar beet. Induction of Na^+/H^+ antiport activity at the tonoplast by growth in salt. Plant Physiol 83: 884–887
Boeshore ML, Hanson MR, Izhar S (1985) A variant mitochondrial DNA arrangement specific to *Petunia* stable sterile somatic hybrids. Plant Mol Biol 4: 125–132
Bouharmont J, Dekeyser A (1989) In vitro selection for cold and salt tolerance in rice. In: Mujeeb-Kazi A, Stich LA (eds) Review of advances in plant biotechnology, 1985–1988: 2nd Int Symp Genetic Manipulation in Crops. CIMMYT and IRRI, Mexico city, and Manila, pp 305–313
Bradford KJ, Hsiao TC (1982) Physiological responses to moderate water stress. In: Lange OL, Nobel PS, Osmond CB, Ziegler H (eds) Physiological plant ecology II. Water relations and carbon assimilation Encyclopedia Plant Physiology (New Series), vol 12B. Springer, Berlin Heidelberg New York, pp 263–324
Braun Y, Hassidim M, Lerner HR, Reinhold L (1986) Studies on H^+-translocating ATPase in plants of varying resistance to salinity. I. Salinity during growth modulates the proton pump in the halophyte *Atriplex nummularia*. Plant Physiol 81: 1050–1056
Braun Y, Hassidim M, Cooper S, Lerner HR, Reinhold L (1989) Are there separate Na^+/H^+ and K^+/H^+ antiporters? In: Dainty J, De Michelis MI, Marre E, Rasi-Caldogno F (eds) Plant membrane transport: the current position. Elsevier, Amsterdam, pp 667–670
Breiman A, Galun E (1990) Nuclear-mitochondrial interrelation in angiosperms. Plant Sci 71: 3–19

Bressan RA, Handa AK, Handa S, Hasegawa PM (1982) Growth and water relations of cultured tomato cells after adjustment to low external water potentials. Plant Physiol 70: 1303–1309
Bressan RA, Singh NK, Handa AK, Kononowicz AK, Hasegawa PM (1985) Stable and unstable tolerance to NaCl in cultured tobacco cells. In: Freeling M (ed) Plant genetics. Liss, New York, pp 755–769
Bressan RA, Singh NK, Handa AK, Mount R, Clithero J, Hasegawa PM (1987) Stability of altered genetic expression in cultured plant cells adapted to salt. In: Monti L, Proceddu E (eds) Drought resistance in plants. Physiological and genetic aspects. Commission of the European Communities, Brussels, pp 41–57
Bressan RA, Nelson DE, Iraki NM, LaRosa PC, Singh NK, Hasegawa PM, Carpita NC (1990) Reduced cell expansion and changes in cell walls of plant cells adapted to NaCl. In: Kattermann F (ed) Environmental injury to plants. Academic Press, New York, pp 137–171
Brettell RS, Dennis ES (1991) Reactivation of a silent *Ac* following tissue culture is associated with heritable alterations in its methylation pattern. Mol Gen Genet 229: 365–372
Briskin DP (1990) The plasma membrane H^+-ATPase of higher plant cells: biochemistry and transport function. Biochim Biophys Acta 1019: 95–109
Broertjes C, van Harten AM (1978) Application of mutation breeding methods in the improvement of vegetatively propagated crops. Elsevier, Amsterdam
Broglie K, Chet I, Holliday M, Cressman R, Biddle P, Knowlton S, Mauvais CJ, Broglie R (1991) Transgenic plants with enhanced resistance to the fungal pathogen *Rhizoctonia solani*. Science 254: 1194–1197
Brouquisse R, Weigel P, Rhodes D, Yocum CF, Hanson AD (1989) Evidence for a ferredoxin-dependent choline monooxygenase from spinach chloroplast stroma. Plant Physiol 90: 322–329
Brunk DG, Rich PJ, Rhodes D (1989) Genotypic variation for glycinebetaine among public inbreds of maize. Plant Physiol 91: 1122–1125
Caboche M (1990) Liposome-mediated transfer of nucleic acids in plant protoplasts. Physiol Plant 79: 173–176
Cia T, Ejeta G, Axtel JD, Botler LG (1990) Somaclonal variation in high tannin sorghums. Theor Appl Genet 79: 737–747
Carlson PR Jr, Forrest J (1982) Uptake of dissolve sulphide by *Spartina alterniflora*: evidence from natural sulfur isotope abundance ratios. Science 216: 633–635
Casas AM, Bressan RA, Hasegawa PM (1991) Cell growth and water relations of the halophyte, *Atriplex nummularia* L, in response to NaCl. Plant Cell Rep 10: 81–84
Chandler SF, Vasil IK (1984) Selection and characterization of NaCl-tolerant cells from embryogenic cultures of *Pennisetum purpureum* Schum. (Napier grass). Plant Sci Lett 37: 157–164
Cheeseman JM (1986) Compartmental efflux analysis: an evaluation of the technique and its limitations. Plant Physiol 80: 1006–1011
Chen Q, Boss WF (1991) Neomycin inhibits the phosphatidylinositol monophosphate and phosphatidylinositol bisphosphate stimulation of plasma membrane ATPase activity. Plant Physiol 96: 340–343
Chitlaru E, Pick U (1991) Regulation of glycerol synthesis in response to osmotic changes in *Dunaliella*. Plant Physiol 96: 50–60
Christou P (1990) Soybean transformation by electric discharge particle acceleration. Physiol Plant 79: 210–212
Corbin DR, Klee HJ (1991) *Agrobacterium tumefaciens*-mediated plant transformation systems. Curr Opin Biotechnol 2: 147–152
Cramer GR, Läuchli A, Polito V (1985) Displacement of Ca^{2+} by Na^+ from the plasmalemma of root cells. A primary response to salt stress? Plant Physiol 79: 207–211
Croughan TP, Stavarek SJ, Rains DW (1978) Selection of a NaCl-tolerant line of cultured alfalfa cells. Crop Sci 18: 959–963

Csonka LN (1989) Physiological and genetic responses of bacteria to osmotic stress. Microbiol Rev 53: 121–147

Cullis CA, Cleary W (1985) Fluidity of the flax genome. In: Freeling M (ed) Plant genetics. Liss, New York, pp 303–310

Cushman JC, Meyer G, Michalowski CB, Schmitt JM, Bohnert HJ (1989) Salt stress leads to differential expression of two isogenes of phosphoenolpyruvate carboxylase during Crassulacean acid metabolism induction in the common ice plant. Plant Cell 1: 715–725

Cushman JC, DeRocher EJ, Bohnert HJ (1990) Gene expression during adaptation to salt stress. In: Kattermann R (ed) Environmental injury to plants. Academic Press, New York, pp 173–203

Cutler JM, Shahan KW, Steponkus (1980) Influence of water deficits and osmotic adjustment on leaf elongation in rice. Crop Sci 20: 314–318

Dacey JWH, King GM, Wakeman SG (1987) Factors controlling emission of dimethylsulphide from salt marshes. Nature 330: 643–645

D'Amato F (1977) Nuclear cytology in relation to development. In: Abercrombie M, Newth DR, Torrey JG (eds) Developmental and cell biology series, 6. Cambridge University Press, London, p 283

D'Amato F (1985) Cytogenetics of plant cell and tissue cultures and their regenerants. CRC Crit Rev Plant Sci 3: 73–112

D'Amato F (1990) Somatic nuclear mutations in vivo and in vitro in higher plants. Caryologia 43: 191–204

Delauney AJ, Verma DPS (1990) A soybean gene encoding Δ^1-pyrroline-5-carboxylate reductase was isolated by functional complementation in *Escherichia coli* and is found to be osmoregulated. Mol Gen Genet 221: 299–305

Dix PJ, Street HE (1975) Sodium chloride-resistant cultured cell lines from *Nicotiana sylvestris* and *Capsicum annuum*. Plant Sci Lett 5: 231–237

Donn G, Tischer E, Smith JA, Goodman HM (1984) Herbicide-resistant alfalfa cells: an example of gene amplification in plants. J Mol Appl Genet 2: 621–635

Dracup M (1991) Increasing salt tolerance of plants through cell culture requires greater understanding of tolerance mechanisms. Aust J Plant Physiol 18: 1–15

Drake BG, Gallagher JL (1984) Osmotic potential and turgor maintenance in *Spartina alterniflora* Loisel. Oecologia 62: 368–375

DuPont FM, Bush DS, Windle JJ, Jones RL (1990) Calcium and proton transport in membrane vesicles from barley roots. Plant Physiol 94: 179–188

Eigel L, Oelmuller R, Koop H-U (1991) Transfer of defined number of chloroplasts into albino protoplasts using an improved subprotoplast/protoplast microfusion procedure: transfer of only two chloroplast leads to variegated progeny. Mol Gen Genet 227: 446–451

Epstein E (1980) Responses of plants to saline environments. In: Rains DW, Valentine RC, Hollaender A (eds) Genetic engineering of osmoregulation. Plenum Press, New York, pp 7–22

Evans DE, Briars S-A, Williams LE (1991) Active calcium transport by plant cell membranes. J Exp Bot 42: 285–303

Fan TW-M, Higashi RM, Norlyn J, Epstein E (1989) In vivo ^{23}Na and ^{31}P NMR measurement of tonoplast Na^+/H^+ exchange process and its distribution characteristics in two barley cultivars. Proc Natl Acad Sci USA 86: 9856–9860

Finch RP, Slamet IH, Cocking EC (1990) Production of heterokaryons by the fusion of mesophyll protoplasts of *Porteresia coarctata* and cell suspension-derived protoplasts of *Oryza sativa*: a new approach to somatic hybridization of rice. J Plant Physiol 136: 592–598

Fischhoff DA, Bowdish KS, Perlak FJ, Marrone PG, McCormick SM, Niedermeyer JG, Dean DA, Kusano-Kreetzmer K, Mayer EJ, Rochester DE, Rogers SG Fraley RT (1987) Insect-tolerant transgenic tomato plants. Bio/Technology 5: 807–813

Flowers TJ, Troke PF, Yeo AR (1977) The mechanism of salt tolerance in halophytes. Annu Rev Plant Physiol 28: 89–121

Flowers TJ, Lachno DR, Flowers SA, Yeo AR (1985) Some effects of sodium chloride on cells of rice cultured in vitro. Plant Sci Lett 39: 205–211
Flowers TJ, Hajibagheri MA, Clipson NJW (1986) Halophytes. Q Rev Biol 61: 313–337
Freytag AH, Wrather JA, Erichsen AW (1990) Salt-tolerant sugarbeet progeny from tissue cultures challenged with multiple salts. Plant Cell Rep 9: 647–650
Fritze K, Staiger D, Czaja I, Walden R, Schell J, Wing D (1991) Developmental and UV light regulation of the snapdragon chalcone synthase promoter. Plant Cell 3: 893–905
Gal S, Pisan B, Hohn T, Grimsley N, Hohn B (1991) Genomic homologous recombination in planta. EMBO J 10: 1571–1578
Gale J, Boll WG (1978) Growth of bean cells in suspension culture in the presence of NaCl and protein-stabilizing factors. Can J Bot 57: 777–782
Galun E, Aviv D (1986) Organelle transfer. Meth Enzymol 118: 595–611
Galun E, Perl A, Aviv D (1988) Protoplast-fusion-mediated transfer of male-sterility and other plastome-controlled traits. In: Application of plant cell and tissue culture. Ciba Foundation Symp 137. Wiley, Chichester, pp 97–112
Garbarino J, DuPont F (1988) NaCl induces a Na^+/H^+ antiport in tonoplast vesicles from barley roots. Plant Physiol 86: 231–236
Garcia-Rios MG, LaRosa PC, Bressan RA, Csonka LN, Hanquier JM (1991) Cloning by complementation of the γ-glutamyl kinase gene from a tomato expression library. Int Soc Plant Molecular Biology 3rd Int Congr, Tucson, Arizona, October 6–11. Abstract 1507
Giannini JL, Gildensoph LH, Reynolds-Niessman I, Briskin DP (1987) Calcium transport in sealed vesicles from red beet (*Beta vulgaris* L.) storage tissue. I. Characterization of a Ca^{2+} pumping ATPase associated with the endoplasmic reticulum. Plant Physiol 85: 1129–1136
Gilmartin PM, Sarokin L, Memelink J, Chua N-H (1990) Molecular light switches for plant genes. Plant Cell 2: 369–378
Gleba YY, Sytnik KM (1984) Protoplast fusion: genetic engineering in higher plants. Springer, Berlin Heidelberg New York
Glenn EP, O'Leary JW (1984) Relationship between salt accumulation and water content of dicotyledonous halophytes. Plant Cell Environ 7: 253–261
Glimelius K (1988) Potentials of protoplast fusion in plant breeding programmes. In: Puite KJ, Dons JJM, Huizing HJ, Kool AJ, Koornneef M, Krens FA (eds) Progress in plant protoplast research. Kluwer Academic, Norwell, pp 159–168
Gobel E, Lorz H (1988) Genetic manipulation of cereals. Oxford Surv Plant Mol Cell Biol 5: 1–22
Goldner R, Umiel N, Chen Y (1977) The growth of carrot callus cultures in various concentrations and compositions of saline water. Z. Pflanzenphysiol 85: 307–317
Gorham J, Wyn Jones RG, Bristol A (1990) Partial characterization of the trait for enhanced K^+–Na^+ discrimination in the D genome of wheat. Planta 180: 590–597
Greef W de, Delon R, Block M de, Leemans J, Botterman J (1989) Evaluation of herbicide resistance in transgenic crops under field conditions. Bio/Technology 7: 61–64
Greenway H, Munns R (1980) Mechanisms of salt tolerance in nonhalophytes. Annu Rev Plant Physiol 31: 149–190
Grumet R, Hanson AD (1986) Genetic evidence for an osmoregulatory function of glycinebetaine accumulation in barley. Aust J Plant Physiol 13: 353–364
Hageman J, Baecke C, Ebskamp M, Pilon R, Smeekens S, Weisbeek P (1990) Protein import into and sorting inside the chloroplast are independent processes. Plant Cell 2: 479–494
Hajibagheri MA, Harvey DMR, Flowers TJ (1987) Quantitative ion distribution within root cells of salt-sensitive and salt-tolerant maize varieties. New Phytol 105: 367–379
Handa AK, Bressan RA, Handa S, Hasegawa PM (1983) Clonal variation for tolerance to polyethylene glycol-induced water stress in cultured tomato cells. Plant Physiol 72: 645–653
Handa S, Handa AK, Hasegawa PM, Bressan RA (1986) Proline accumulation and the adaptation of cultured plant cells to water stress. Plant Physiol 80: 938–945

Hanning G, Nabors M (1989) In vitro tissue culture selection for sodium chloride (NaCl) tolerance in rice and the performance of the regenerants under saline conditions. In: Mujeeb-Kazi A, Stich LA (eds) Review of advances in plant biotechnology, 1985–1988. 2nd Int Symp Genetic Manipulation in Crops. CIMMYT and IRRI, Mexico city, and Manila, pp 239–248
Hanson AD, Hitz WD (1982) Metabolic responses of mesophytes to plant water deficits. Annu Rev Plant Physiol 33: 163–203
Hanson AD, Rhodes D (1983) ^{14}C Tracer evidence for synthesis of choline and betaine via phosphoryl base intermediates in salinized sugar beet leaves. Plant Physiol 71: 692–700
Hanson AD, May AM, Grumet R, Bode J, Jamieson GC, Rhodes D (1985) Betaine synthesis in chenopods: localization in chloroplasts. Proc Natl Acad Sci USA 82: 3678–3682
Hanson AD, Rathinasabapathi B, Chamberlin B, Gage DA (1991) Comparative physiological evidence that β-alanine betaine and choline-O-sulphate act as compatible osmolytes in halophytic *Limonium* species. Plant Physiol 97: 1199–1205
Hanson Mr (1984) Cell culture and recombinant DNA methods for understanding and improving salt resistance of plants. In: Staples RC, Toenniessen GH (eds) Salinity tolerance in plants. Strategies for crop improvement. Wiley, New York, pp 335–359
Harms CT, Oertli JJ (1985) The use of osmotically adapted cell cultures to study salt tolerance in vitro. J Plant Physiol 120: 29–38
Hasegawa PM, Bressan RA, Handa AK (1980) Growth characteristics of NaCl-selected and nonselected cells of *Nicotiana tabacum* L. Plant Cell Physiol 21: 1347–1355
Hasegawa PM, Bressan RA, Handa AK (1986) Cellular mechanisms of salinity tolerance. HortScience 21: 1317–1324
Hasegawa PM, Binzel ML, Reuveni M, Watad AA, Bressan RA (1990) Physiological and molecular mechanisms of ion accumulation and compartmentation contributing to salt adaptation of plant cells. In: Bennett AB, O'Neill SD (eds) Horticulture biotechnology. Wiley-Liss, New York, pp 295–304
Hassidim M, Braun Y, Lerner HR, Reinhold L (1990) Na^+/H^+ and K^+/H^+ antiport in root membrane vesicles isolated from the halophyte *Atriplex* and the glycophyte cotton. Plant Physiol 95: 1795–1801
Hedrich R, Schroeder JI (1989) The physiology of ion channels and electrogenic pumps in higher plants. Annu Rev Plant Physiol Plant Mol Biol 40: 539–569
Heyser JW, Nabors MW (1981a) Osmotic adjustment of cultured tobacco cells (*Nicotiana tabacum* var. Samsun) grown on sodium chloride. Plant Physiol 67: 720–727
Heyser JW, Nabors MW (1981b) Growth, water content and solute accumulation of two cell lines cultured on sodium chloride, dextran and polyethylene glycol. Plant Physiol 68: 1454–1459
Hitz WD, Hanson AD (1980) Determination of glycine betaine by pyrolysis-gas chromatography in cereals and grasses. Phytochemistry 19: 2371-2374
Hitz WD, Rhodes D, Hanson AD (1981) Radiotracer evidence implicating phosphoryl and phosphatidyl bases as intermediates in betaine synthesis by water-stressed barley leaves. Plant Physiol 68: 814–822
Hsiao TC, Jing J (1987) Leaf and root expansive growth in response to water deficits. In: Cosgrove DJ, Knievel DP (eds) Physiology of plant growth in response to water deficits. American Society Plant Physiologists, Rockville, MD, pp 180–192
Incharoensakdi A, Takabe T, Akazawa T (1986) Effect of betaine on enzyme activity and subunit interaction of ribulose-1,5-bisphosphate carboxylase/oxygenase from *Aphanothece halophytica*. Plant Physiol 81: 1044–1049
Jain RK, Jain S, Nainawatee HS, Chowdhury JB (1990) Salt tolerance in *Brassica juncea* L. I. In vitro selection, agronomic evaluation and genetic stability. Euphytica 48: 141–152
Jain RK, Jain S, Chowdhury JB (1991) In vitro selection for salt tolerance in *Brassica juncea* L. Using cotyledon explants, callus and cell suspension cultures. Ann Bot 67: 517–519

Jeschke WD (1984) K^+–Na^+ exchange at cellular membranes, intracellular compartmentation of cations, and salt tolerance. In: Staples RC, Toennissen GH (eds) Salinity tolerance in plants. Strategies for crop improvement. Wiley, New York, pp 37–66

Kacser H (1983) The control of enzyme systems in vivo: elasticity analysis of the steady state. Biochem Soc Trans 11: 35–40

Kaniewski W, Lawson C, Sammons B, Haley L, Hart J, Delannay X, Tumer NE (1990) Field resistance of transgenic russet burbank potato to effects of infection by potato virus X and potato virus Y. Bio/Technology 8: 750–754

Karp A (1991) On the current understanding of somaclonal variation. In: Miflin BJ, Miflin HF (eds) Oxford surveys of plant molecular and cell biology. Oxford University Press, Oxford, pp 1–58

Kauss H (1981) Sensing of volume changes by *Pterioochromonas* involves a Ca^{2+}-regulated system which controls activation of isofloridoside phosphate synthase. Plant Physiol 68: 420–424

Kauss H (1983) Volume regulation in *Pterioochromonas*. Involvement of calmodulin in the Ca^{2+}-stimulated activation of isofloridoside phosphate synthase. Plant Physiol 71: 169–172

Kauss H (1987) Some aspects of calcium-dependent regulation in plant metabolism. Annu Rev Plant Physiol 38: 47–72

Kochba J, Ben-Hayyim G, Spiegel-Roy P, Saad S, Neumann H (1982) Selection of stable salt-tolerant callus cell lines and embryos in *Citrus sinensis* and *C. aurantium*. Z Pflanzenphysiol 106: 111–118

Koltunow AM, Truettner J, Cox KH, Wallroth M, Goldberg RB (1990) Different temporal and spatial gene expression patterns occur during anther development. Plant Cell 2: 1201–1224

Kononowicz AK, Floryanowicz-Czekalska K, Clithero J, Meyers A, Hasegawa PM, Bressan RA (1990a) Chromosome number and DNA content of tobacco cells adapted to NaCl. Plant Cell Rep 9: 672–675

Kononowicz AK, Hasegawa PM, Bressan RA (1990b) Chromosome number and nuclear DNA content of plants regenerated from salt-adapted plant cells. Plant Cell Rep 9: 676–679

Kriedemann PE (1986) Stomatal and photosynthetic limitations to leaf growth. Aust J Plant Physiol 13: 15–31

Larkin PJ, Scowcroft WR (1981) Somaclonal variation – a novel source of variability from cell cultures for plant improvement. Theor Appl Genet 60: 197–214

LaRosa PC, Handa AK, Hasegawa PM, Bressan RA (1985) Abscisic acid accelerates adaptation of cultured tobacco cells to salt. Plant Physiol 79: 138–142

LaRosa PC, Hasegawa PM, Rhodes D, Clithero JM, Watad AA, Bressan RA (1987) Abscisic acid stimulated osmotic adjustment and its involvement in adaptation of tobacco cells to salt. Plant Physiol 85: 174–181

LaRosa PC, Singh NK, Hasegawa PM, Bressan RA (1989) Stable NaCl tolerance of tobacco cells is associated with enhanced accumulation of osmotin. Plant Physiol 91: 855–861

LaRosa PC, Rhodes D, Rhodes JC, Bressan RA, Csonka LN (1991) Elevated accumulation of proline in NaCl-adapted tobacco cells is not due to altered Δ^1-pyrroline-5-carboxylate reductase. Plant Physiol 96: 245–250

Läuchli A (1990) Calcium, salinity and the plasma membrane. In: Leonard RT, Hepler PK (eds) Calcium in plant growth and development. Am Soc Plant Physiol, Rockville, MD, pp 26–35

Läuchli A, Epstein E (1970) Transport of potassium and rubidium in plant roots. The significance of calcium. Plant Physiol 45: 639–641

Lee KY, Lund P, Lowe K, Dunsmuir P (1990) Homologous recombination in plant cells after *Agrobacterium*-mediated transformation. Plant Cell 2: 415–425

Lee M, Phillips RL (1988) The chromosomal basis of somaclonal variation. Annu Rev Plant Physiol Plant Mol Biol 39: 413–437

Lerma C, Hanson AD, Rhodes D (1988) Oxygen-18 and deuterium labeling studies of choline oxidation by spinach and sugar beet. Plant Physiol 88: 695–702

Lerma C, Rich PJ, Ju GC, Yang W-J, Hanson AD, Rhodes D (1991) Betaine deficiency in maize: complementation tests and metabolic basis. Plant Physiol 95: 1113–1119
Lindsey K, Jones MKG (1990) Electroporation of cells. Physiol Plant 79: 168–172
Manetas Y (1990) A re-examination of NaCl effects on phosphoenolpyruvate carboxylase at high (physiological) enzyme concentrations. Physiol Plant 78: 225–229
Mariani C, De Beuckeleer M, Turettner J, Leemans J, Goldberg RB (1990) Induction of male sterility in plants by a chimaeric ribonuclease gene. Nature 347: 737–741
Mathur AK, Ganapathy PS, Johri BM (1980) Isolation of sodium chloride-tolerant plantlets of *Kichxia ramosissima* under in vitro conditions. Z Pflanzenphysiol 99: 287–294
Matsuda K, Riazi A (1981) Stress-induced osmotic adjustment in growing regions of barley leaves. Plant Physiol 68: 571–576
McCoy TJ (1987a) Tissue culture evaluation of NaCl tolerance in *Medicago* species: cellular versus whole plant response. Plant Cell Rep 6: 31–34
McCoy TJ (1987b) Characterization of alfalfa (*Medicago sativa* L.) plants regenerated from NaCl-tolerant cell lines. Plant Cell Rep 6 : 417–422
McCue KF, Hanson AD (1990) Drought and salt tolerance: towards understanding and application. Trends Biotechnol 8: 358–362
McHughen AG (1987) Salt tolerance through increased vigor in a flax line (STS-11) selected for salt tolerance in vitro. Theor Appl Genet 74: 727–732
Medgysey P, Fejes E, Maliga P (1985) Interspecific chloroplast recombination in a *Nicotiana* somatic hybrid. Proc Natl Acad Sci USA 82: 6960–6964
Meins F Jr (1983) Heritable variation in plant cell culture. Annu Rev Plant Physiol 34: 327–346
Meyer RF, Boyer JS (1981) Osmoregulation, solute distribution and growth in soybean seedlings having low water potentials. Planta 151: 482–489
Michelena VA, Boyer JS (1982) Complete turgor maintenance at low water potentials in the elongating region of maize leaves. Plant Physiol 69: 1145–1149
Moll B, Polsby L, Maliga P (1990) Streptomycin and lincomycin resistances are selective markers in cultured *Nicotiana* cells. Mol Gen Genet 221: 245–250
Morgan A, Maliga P (1987) Rapid chloroplast segregation and recombination of mitochondrial DNA in *B. napus* cybrids. Mol Gen Genet 209: 240–246
Munns R, Termaat A (1986) Whole-plant responses to salinity. Aust J Plant Physiol 13: 143–160
Nabors MW (1976) Using spontaneously occurring and induced mutations to obtain agriculturally useful plants. BioScience 26: 761–768
Nabors MW (1983) Increasing the salt and drought tolerance of crop plants. In: Randall DD, Blevins DG, Larson RL, Rapp BJ (eds) Current topics in plant biochemistry and physiology, vol 2. University of Missouri, Columbia, pp 165–184
Nabors MW (1990) Environmental stress resistance. In: Dix PJ (ed) Plant cell line selection. Procedures and applications. VCH, Weinheim, pp 167–186
Narbors MW, Dykes TA (1985) Tissue culture of cereal cultivars with increased salt, drought and acid tolerance. In: Biotechnology in international agricultural research. International Rice Research Institute, Manila, Philippines, pp 121–138
Nabors MW, Daniels A, Nadolny L, Brown C (1975) Sodium chloride-tolerant lines of tobacco cells. Plant Sci Lett 4: 155–159
Nabors MW, Gibbs SW, Bernstein CS, Meins ME (1980) NaCl-tolerant tobacco plants from cultured cells. Z Pflanzenphysiol 97: 13–17
Nabors MW, Kroskey CS, McHugh DM (1982) Green spots are predictors of high callus growth rates and shoot formation in normal and salt stressed tissue cultures of oat (*Avena sativa* L.). Z. Pflanzenphysiol 105: 341–349
Napoli C, Lemieux C, Jorgensen R (1990) Introduction of a chimeric chalcone synthase gene into petunia results in reversible co-suppression of homologous genes in trans. Plant Cell 2: 279–289
Narasimhan ML, Binzel ML, Perez-Prat E, Chen Z, Nelson DE, Singh NK, Bressan RA, Hasegawa PM (1991) NaCl regulation of tonoplast ATPase 70-kilodalton subunit mRNA in tobacco cells. Plant Physiol 97: 562–568

Narayanan KK, Sree Rangasamy SR (1989) Inheritance of salt tolerance in progenies of tissue culture selected variants of rice. Curr Sci 58: 1204–1205
Neuhaus G, Spangenberg G (1990) Plant transformation by microinjection techniques. Physiol Plant 79: 213–217
Oard JH (1991) Physical methods for the transformation of plant cells. Biotechnol Adv 9: 1–11
Ochatt SJ, Power JB (1988) Selection of salt and drought tolerance in protoplast-and explant-derived tissue cultures of colt cherry (*Prunus avium×Pseudocerasus*). Tree Physiol 5: 259–266
Oeller PW, Min-Wong L, Taylor LP, Pike DA, Theologis A (1991) Reversible inhibition of tomato fruit senescence by antisense RNA. Science 254: 437–439
Orton TJ (1980) Comparison of salt tolerance between *Hordeum vulgare* and *H. jubatum* in whole plants and callus cultures. Z Pflanzenphysiol 98: 105–118
Pahlich E, Kerres R, Jager H-J (1983) Influence of water stress on the vacuole/extra-vacuole distribution of proline in protoplasts of *Nicotiana rustica*. Plant Physiol 72: 590–591
Paleg LG, Stewart GR, Bradbeer JW (1984) Proline and glycine betaine influence protein solvation. Plant Physiol 75: 974–978
Pallaghy CK, Scott BIH (1969) The electrochemical state of cells of broad bean roots II. Potassium kinetics in excised root tissue. Aust J Biol Sci 22: 585–600
Partanen C (1965) On the chromosomal basis for cellular differentiation. Am J Bot 52: 204–209
Paszkowski J, Baur M, Bogucki A, Potrykus I (1988) Gene targeting in plants. EMBO J 7: 4021–4026
Perez-Prat E, Binzel ML, Bressan RA, Valpuesta V, Hasegawa PM (1990) Isolation and characterization of a cDNA clone encoding an E_1E_2 type ATPases from *Nicotiana tabacum*. Plant Physiol 93 (Suppl): 104 (Abstr)
Perl A, Aviv D, Galun E (1991) Protoplast fusion mediated transfer of oligomycin resistance from *Nicotiana sylvestris* to *Solanum tuberosum* by intergeneric cybridization. Mol Gen Genet 225: 11–16
Pesci P (1987) ABA-induced proline accumulation in barley leaf segments: dependence on protein synthesis. Physiol Plant 71: 287–291
Pesci P (1989) Involvement of Cl^- in the increase in proline induced by ABA and stimulated by potassium chloride in barley leaf segments. Plant Physiol 89: 1226–1230
Pesci P, Beffagna N (1985) Effects of weak acids on proline accumulation in barley leaves: a comparison between abscisic acid and isobutyric acid. Plant Cell Environ 8: 129–133
Phillips RL, Kaeppler SM, Peschke VM (1990) Do we understand somaclonal variation? In: Nijkamp HJJ, Van der Plas LHW, Van Aartrijk J (eds) Progress in plant cellular and molecular biology. Kluwer, Dordrecht, pp 131–141
Poljakoff-Mayber A (1982) Biochemical and physiological responses of higher plants to salinity stress. In: San Pietro A (ed) Biosaline research, a look to the future. Plenum Press, New York, pp 245–269
Poljakoff-Mayber A, Symon DE, Jones GP, Naidu BP, Paleg LG (1987) Nitrogenous compatible solutes in native South Australian plants. Aust J Plant Physiol 14: 341–350
Pollard A, Wyn Jones RG (1979) Enzyme activities in concentrated solutions of glycine betaine and other solutes. Planta 144: 291–298
Poole RJ (1988) Plasma membrane and tonoplast. In: Baker DA, Hall JL (eds) Solute transport in plant cells and tissues. Wiley, New York, pp 83–105
Potrykus I (1990) Gene transfer to cereals: an assessment. Bio/Technology 8: 535–542
Potrykus I (1991) Gene transfer to plants: assessment of published approaches and results. Annu Rev Plant Physiol Plant Mol Biol 42: 205–225
Pua EC, Thorpe TA (1986) Differential Na_2SO_4 tolerance in tobacco plants regenerated from Na_2SO_4-grown callus. Plant Cell Environ 9: 9–16
Raghava Ram NV, Nabors MW (1985) Salinity tolerance. In: Cheremisinoff PN, Quellette RP (eds) Biotechnology: applications and research. Technomic, Lancaster, pp 623–642

Rains DW (1972) Salt transport by plants in relation to salinity. Annu Rev Plant Physiol 23: 367–388
Ramage RT (1980) Genetic methods to breed salt tolerance in plants. In: Rains DW, Valentine RC, Hollaender A (eds) Genetic engineering of osmoregulation. Impact on plant productivity for food, chemicals and energy. Plenum Press, New York, pp 311–318
Rasi-Caldogno F, Pugliarello MC, De Michelis MI (1987) The Ca^{2+}-transport ATPase of plant plasma membrane catalyzes a nH^+/Ca^{2+} exchange. Plant Physiol 83: 994–1000
Rea PA, Sanders D (1987) Tonoplast energization: two H^+ pumps, one membrane. Physiol Plant 81: 131–141
Reinert J (1958a) Untersuchungen über die Morphogenese in Gewebekulturen. Ber Dtsch Bot Ges 71: 15
Reinert J (1958b) Morphogenese und ihre Kontrolle an Gewebekulturen aus karotten. Naturwissenschaften 45: 244–245
Reinert J (1959) Über die Kontrolle der Morphogenese und die Induktion von Adventivembryonen in Gewebekulturen aus Karotten. Planta 53: 318–338
Reinhold L, Kaplan A (1984) Membrane transport of sugars and amino acids. Annu Rev Plant Physiol 35: 45–83
Reinhold L, Seiden A, Volikita M (1984) Is modulation of the rate of proton pumping a key event in osmoregulation? Plant Physiol 75: 846–849
Reinhold L, Braun Y, Hassidim M, Lerner HR (1989) The possible role of various membrane transport mechanisms in adaptation to salinity. In: Cherry JH (ed) Biochemical and physiological mechanisms associated with environmental stress tolerance. Springer, Berlin Heidelberg New York, pp 121–130
Reuveni M, Bennett AB, Bressan RA, Hasegawa PM (1990) Enhanced H^+ transport capacity and ATP hydrolysis activity of the tonoplast H^+-ATPase after NaCl adaptation. Plant Physiol 94: 524–530
Rhodes D (1987) Metabolic responses to stress. In: Davis DD (ed) The biochemistry of plants, vol 12. Academic Press, New York, pp 201–241
Rhodes D, Handa S (1989) Amino acid metabolism in relation to osmotic adjustment in plant cells. In: Cherry JH (ed) Biochemical and physiological mechanisms associated with environmental stress tolerance. Springer, Berlin Heidelberg New York, pp 41–62
Rhodes D, Rich PJ (1988) Preliminary genetic studies of the phenotype of betaine deficiency in *Zea mays* L. Plant Physiol 88: 102–108
Rhodes D, Handa S, Bressan RA (1986) Metabolic changes associated with adaptation of plant cells to water stress. Plant Physiol 82: 890–903
Rhodes D, Rich PJ, Myers AC, Reuter CR, Jamieson GC (1987) Determination of betaines by fast atom bombardment mass spectrometry: identification of glycine betaine deficient genotypes of *Zea mays*. Plant Physiol 84: 781–788
Rhodes D, Rich PJ, Brunk DG, Ju GC, Rhodes JC, Pauly MH, Hansen LA (1989) Development of two isogenic sweet corn hybrids differing for glycinebetaine content. Plant Physiol 91: 1112–1121
Rietveld RC, Singh NK, Hasegawa PM, Bressan RA (1988) A selectable mtDNA polymorphism is found in salt-tolerant tobacco mitochondria. Plant Physiol (Suppl) 86: 136 (Abstr)
Robinson C, Larsson C, Buckhout TJ (1988) Identification of a calmodulin-stimulated (Ca^{2+} + Mg^{2+})-ATPase in a plasma membrane fraction isolated from maize (*Zea mays*) leaves. Physiol Plant 72: 177–184
Robinson SP, Jones GP (1986) Accumulation of glycine betaine in chloroplasts provides osmotic adjustment during salt stress. Aust J Plant Physiol 13: 659–668
Rogers SG (1991) Free DNA methods for plant transformation. Curr Opin Biotechnol 2: 153–157
Samac DA, Shah DM (1991) Developmental and pathogen-induced activation of the *Arabidopsis* acidic chitinase promoter. Plant Cell 3: 1063–1072

Sanchez-Aguayo I, Gonzalez-Utor AL, Medina A (1991) Cytochemical localization of ATPase activity in salt-treated and salt-free grown *Lycopersicon esculentum* roots. Plant Physiol 96: 153–158

Sanford JC (1990) Biolistic plant transformation. Physiol Plant 79: 206–209

Schachtman DP, Bloom AJ, Dvorak J (1989) Salt-tolerant *Triticum* × *Lophopyrum* derivatives limit the accumulation of sodium and chloride ions under saline stress. Plant Cell Environ 12: 47–55

Schachtman DP, Tyerman SD, Terry BR (1991) The K^+/Na^+ selectivity of a cation channel in the plasma membrane of root cells does not differ in salt-tolerant and salt-sensitive wheat species. Plant Physiol 97: 598–605

Schmid J, Doemer PW, Clouse SD, Dixon RA, Lamb CJ (1990) Developmental and environmental regulation of a bean chalcone synthase promoter in trangenic tobacco. Plant Cell 2: 619–631

Schnapp SR, Bressan RA, Hasegawa PM (1990) Carbon use efficiency and cell expansion of NaCl-adapted tobacco cells. Plant Physiol 93: 384–388

Schnapp SR, Curtis WR, Bressan RA, Hasegawa PM (1991) Growth yields and maintenance coefficients of unadapted and NaCl-adapted tobacco cells grown in semicontinuous culture. Plant Physiol 96: 1289–1293

Schobert B (1977) Is there an osmotic regulatory mechanism in algae and higher plants? J Theor Biol 68: 17–26

Schobert B, Tschesche T (1978) Unusual solution properties of proline and its interaction with proteins. Biochim Biophys Acta 541: 270–277

Schroeder JI, Hagiwara S (1989) Cytosolic calcium regulates ion channels in the plasma membrane of *Vicia faba* guard cells. Nature 338: 427–430

Schroeder JI, Thuleau P (1991) Ca^{2+} channels in higher plants. Plant Cell 3: 555–559

Schroppel-Meier G, Kaiser WM (1988) Ion homeostasis in chloroplasts under salinity and mineral deficiency. I. Solute concentrations in leaves and chloroplasts from spinach plants under NaCl or $NaNO_3$ salinity. Plant Physiol 87: 822–827

Schwab KB, Gaff DF (1990) Influence of compatible solutes on soluble enzymes from desiccation-tolerant *Sporobolus stapfianus* and desiccation-sensitive *Sporobolus pyramidalis*. J Plant Physiol 137: 208–215

Serrano R (1989) Structure and function of plasma membrane ATP. Annu Rev Plant Physiol Plant Mol Biol 40: 61–94

Shah SH, Gorham J, Forster BP, Wyn Jones RG (1987) Salt tolerance in Triticeae: the contribution of the D genome to cation selectivity in hexaploid wheat. J Exp Bot 38: 254–269

Shomer-Ilan A, Waisel Y (1986) Effects of stabilizing solutes on salt activation of phosphoenolpyruvate carboxylase from various plant sources. Physiol Plant 67: 408–411

Shomer-Ilan A, Jones GP, Paleg LG (1991) In vitro thermal and salt stability of pyruvate kinase are increased by proline analogues and trigonelline. Aust J Plant Physiol 18: 279–286

Singh NK, Nelson DE, LaRosa PC, Hasegawa PM, Bressan RA (1989) Osmotin: A protein associated with osmotic stress adaptation in plant cells. In: Cherry JH (ed) Biochemical and physiological mechanisms associated with environmental stress tolerance in plants. Springer, Berlin Heidelberg New York, pp 67–87

Skirvin RM (1978) Natural and induced variation in tissue culture. Euphytica 27: 241–266

Skokut TA, Filner P (1980) Slow adaptive changes in urease levels of tobacco cells cultured on urea and other nitrogen sources. Plant Physiol 65: 995–1003

Skoog FK, Miller CO (1957) Chemical regulation of growth and organ formation in plant tissues cultured in vitro. Symp Soc Exp Biol 11: 118–131

Skriver K, Mundy J (1990) Gene expression in response to abscisic acid and osmotic stress. Plant Cell 2: 503–512

Smith MK, McComb JA (1981a) Effect of NaCl on the growth of whole plants and their corresponding callus cultures. Aust J Plant Physiol 8: 267–275

Smith MK, McComb JA (1981b) Use of callus cultures to detect NaCl tolerance in cultivars of three species of pasture legumes. Aust J Plant Physiol 8: 437–422
Somero GN (1986) Protons, osmolytes, and fitness of internal milieu for protein function. Am J Physiol 251: R197–R213
Spanswick RM (1981) Electrogenic ion pumps. Annu Rev Plant Physiol 32: 267–289
Spiegel-Roy P, Ben-Hayyim G (1985) Selection and breeding for salinity tolerance in vitro. Plant Soil 89: 243–252
Spollen WG, Sharp RE (1991) Spatial distribution of turgor and root growth at low water potentials. Plant Physiol 96: 438–443
Stavarek SJ, Rains DW (1984) Cell culture techniques: selection and physiological studies of salt tolerance. In: Staples RC, Toenniessen GH (eds) Salinity tolerance in plants. Strategies for crop improvement. Wiley, New York, pp 321–334
Stavarek SJ, Rains DW (1985) Effect of salinity on growth and maintenance costs of plant cells. In: Key JL, Kosuge T (eds) Cellular and molecular biology. Liss, New York, pp 129–143
Steudler PA, Peterson BJ (1984) Contribution of gaseous sulphur from salt marshes to the global sulphur cycle. Nature 311: 455–457
Stewart CR (1980) The mechanism of abscisic acid-induced proline accumulation in barley leaves. Plant Physiol 66: 230–233
Stewart CR (1981) Proline accumulation: biochemical aspects. In: Paleg LG, Aspinall D (eds) The physiology and biochemistry of drought resistance in plants. Academic Press, Sydney, pp 243–259
Stewart CR, Voetberg G, Rayapati PJ (1986) The effects of benzyladenine, cycloheximide, and cordycipin on wilting-induced abscisic acid and proline accumulation and abscisic acid- and salt-induced proline accumulation in barley leaves. Plant Physiol 82: 703–707
Steward FC, Mapes MO, Mears K (1958) Growth and organized development of cultured cells. II. Organization in cultures grown from freely suspended cells. Am J Bot 45: 705–708
Stewart GR, Larher F (1980) Accumulation of amino acids and related compounds in relation to environmental stress. In: Miflin BJ (ed) The biochemistry of plants: a comprehensive treatise, vol 5. Academic Press, New York, pp 609–635
Stewart GR, Rhodes D (1978) Nitrogen metabolism of halophytes. III. Enzymes of ammonia assimilation. New Phytol 80: 307–316
Stewart GR, Larher F, Ahmad I, Lee JA (1979) Nitrogen metabolism and salt-tolerance in higher plant halophytes. In: Jefferies RL, Davy AJ (eds) Ecological processes in coastal environments. Blackwell Scientific, Oxford, pp 211–227
Storey R, Pitman MG, Stelzer R, Carter C (1983) X-ray micro-analysis of cells and cell compartments of *Atriplex spongiosa*. J Exp Bot 34: 778–794
Strogonov BP (1973) Structure and function of plant cells in saline habitats. Isr Program Sci Translation. Wiley, New York
Struve I, Lüttge U (1987) Characteristics of $MgATP^{2-}$ dependent electrogenic proton transport in tonoplast vesicles of the facultative Crassulacean acid metabolism plant *Mesembryanthemum crystallinum* L. Planta 170: 111–120
Sundberg E, Lagercrantz U, Glimelius K (1991) Effects of cell type used for fusion on chromosome elimination and chloroplast segregation in *Brassica oleracea* (+) *Brassica napus* hybrids. Plant Sci 78: 89–98
Surowy TK, Boyer JS (1991) Low water potentials affect expression of genes encoding vagetative storage proteins and plasma membrane proton ATPase in soybean. Plant Mol Biol 16: 251–262
Sussman MR, Harper JF (1989) Molecular biology of the plasma membrane of higher plants. Plant Cell 1: 953–960
Svab Z, Hajdukiewicz P, Maliga P (1990) Stable transformation of plastids in higher plants. Proc Natl Acad Sci USA 87: 8526–8530
Sze H (1985) H^+-translocating ATPases: advances using membrane vesicles. Annu Rev Plant Physiol 36: 175–208

Tal M (1984) Physiological genetics of salt resistance in higher plants: studies on the level of the whole plant and isolated organs, tissues and cells. In: Staples RC, Toenniessen GH (eds) Salinity tolerance in plants: strategies for crop improvement. Wiley, New York, pp 301–320

Tal M (1990) Somaclonal variation for salt resistance. In: Bajaj YPS (eds) Biotechnology in agriculture and forestry. Somaclonal variation in crop improvement I. Springer, Berlin Heidelberg New York, pp 236–257

Taylor WC (1989) Regulatory interactions between nuclear and plastid genomes. Annu Rev Plant Physiol Plant Mol Biol 40: 211–233

Termaat A, Passioura JB, Munns R (1985) Shoot turgor does not limit shoot growth of NaCl-affected wheat and barley. Plant Physiol 77: 869–872

Tester M (1990) Plant ion channels: whole-cell and single-cell channel studies. New Phytol 114: 305–340

Torrey JG (1967) Development in flowering plants. Macmillan, New York

Tyagi AK, Rashid A, Maheshwari SC (1981) Sodium chloride-resistant cell line from haploid *Datura innoxia* Mill. A resistance trait carried from cell to plantlet and vice versa in vitro. Protoplasma 105: 327–332

Vajrabhaya M, Thanapaisal T, Vajrabhaya T (1989) Development of salt-tolerant lines KDML and LPT rice cultivars through tissue culture. Plant Cell Rep 8: 411–414

van der Krol AR, Chua N-H (1991) The basic domain of plant B-ZIP proteins facilitates import of a receptor protein into plant nuclei. Plant Cell 3: 667–675

van der Krol AR, Mur LA, Beld M, Mol JNM, Stuitje AR (1990) Flavonoid genes in petunia: addition of a limited number of gene copies may lead to a suppression of gene expression. Plant Cell 2: 291–299

van Swaaij AC, Jacobsen E, Kiel JAKW, Feenstra WJ (1986) Selection, characterization and regeneration of hydroxyproline-resistant cell lines of *Solanum tuberosum*: tolerance to NaCl and freezing stress. Physiol Plant 68: 359–366

Vasil IK (1987) Developing cell and tissue culture systems for the improvement of cereal and grass crops. J Plant Physiol 128: 193–218

Voetberg G, Sharp RE (1991) Growth of the maize primary root at low water potentials. III. Role of increased proline deposition in osmotic adjustment. Plant Physiol 96: 1125–1130

Waara S, Wallin A, Eriksson T (1991) Production and analysis of intraspecific somatic hybrids of potato (*Solanum tuberosum* L.) Plant Sci 75: 107–115

Waisel Y (1972) Biology of halophytes. Academic Press, New York

Walker NA, Pitman MG (1976) Measurement of fluxes across membranes. In: Luttge U, Pitman MG (eds) Encyclopedia of plant physiology, New Series, vol 2A. Springer, Berlin Heidelberg, New York, pp 93–120

Warren RS, Gould AR (1982) Salt tolerance expressed as a cellular trait in suspension cultures developed from the halophytic grass *Distichlis spicata*. Z Pflanzenphysiol 107: 347–356

Waskom RM, Miller DR, Hanning GE, Duncan RR, Voigt RL, Nabors MW (1990) Field evaluation of tissue culture derived sorghum for increased tolerance to acid soils and drought stress. Can J Plant Sci 70: 997–1004

Watad AA, Reinhold L, Lerner HR (1983) Comparison between a stable NaCl-selected *Nicotiana* cell line and the wild type. K^+, Na^+ and proline pools as a function of salinity. Plant Physiol 73: 624–629

Watad AA, Lerner HR, Reinhold L (1985) Stability of salt-resistance character in *Nicotiana* cell lines adapted to grow in high NaCl concentrations. Physiol Veg 23: 887–894

Watad AA, Pesci P-A, Reinhold L, Lerner HR (1986) Proton fluxes as a response to external salinity in wild type and NaCl-adapted *Nicotiana* cell lines. Plant Physiol 81: 454–459

Watad AA, Reuveni M, Bressan RA, Hasegawa PM (1991a) Enhanced net K^+ uptake capacity of NaCl-adapted cells. Plant Physiol 95: 1265–1269

Watad AA, Swartzberg D, Bressan RA Izhar S, Hasegawa PM (1991b) Stability of salt tolerance at the cell level after regeneration of plants from salt-tolerant tobacco cell line. Physiol Plant 83: 307–313
Wenzel G, Schieder O, Przewozny T, Sopory SK, Melchers G (1979) Comparison of single cell culture-derived *Solanum tuberosum* L. plants and a model for their application in breeding programs. Theor Appl Genet 55: 49–55
Weretilnyk EA, Hanson AD (1989) Biochemical and genetic characterization of betaine aldehyde dehydrogenase. In: Cherry JH (ed) Biochemical and physiological mechanisms associated with environmental stress tolerance. Springer, Berlin Heidelberg New York, p 65
Weretilnyk EA, Hanson AD (1990) Molecular cloning of a plant betaine-aldehyde dehydrogenase, an enzyme implicated in adaptation to salinity and drought. Proc Natl Acad Sci USA 87: 2745–2749
Weretilnyk EA, Bednarek S, McCue KF, Rhodes D, Hanson AD (1989) Comparative biochemical and immunological studies of the glycine betaine synthesis pathway in diverse families of dicotyledons. Planta 178: 342–352
Wijbrandi J, Posthuma A, Kok JM, Rijken R, Vos JGM, Koornneef M (1990a) Asymmetric somatic hybrids between *Lycopersicon esculentum* and irradiated *Lycopersicon peruvianum*. 1. Cytogenetics and morphology. Theor Appl Genet 80: 305–312
Wijbrandi J, Wolters AMA, Koornneef M (1990b) Asymmetric somatic hybrids between *Lycopersicon esculentum* and irradiated *Lycopersicon peruvianum*. 2. Analysis with marker genes. Theor Appl Genet 80: 665–672
Wijbrandi J, Zabel P, Koornneef M (1990c) Restriction fragment length polymorphism analysis of somatic hybrids between *Lycopersicon esculentum* and irradiated *L. peruvianum*: evidence for limited donor genome elimination and extensive chromosome rearrangements. Mol Gen Genet 222: 270–277
Winicov I (1991) Characterization of salt tolerant alfalfa (*Medicago sativa* L.) plants regenerated from salt tolerant cell lines. Plant Cell Rep 10: 561–564
Wolf O, Munns R, Tonnet ML, Jeschke WD (1990) Concentration and transport of solutes in xylem and phloem along the leaf axis of NaCl-treated *Hordeum vulgare*. J Exp Bot 41: 1133–1141
Wolf O, Munns R, Tonnet ML, Jeschke WD (1991) The role of the stem in the partitioning of Na^+ and K^+ in salt-treated barley. J Exp Bot 42: 697–704
Wyn Jones RG (1981) Salt tolerance. In: Johnson EB (ed) Physiological processes limiting plant productivity. Butterworth, London, pp 271–292
Wyn Jones RG (1984) Phytochemical aspects of osmotic adaptation. Rec Adv Phytochem 18: 55–78
Wyn Jones RG, Gorham J (1983) Aspects of salt and drought tolerance in higher plants. In: Kosuge T, Meredith CP, Hollaender A (eds) Genetic engineering of plants, an agricultural perspective. Plenum Press, New York, pp 355–370
Wyn Jones RG, Storey R (1981) Betaines. In: Paleg LG, Aspinall D (eds) The physiology and biochemistry of drought resistance in plants. Academic Press, Sydney, pp 171–204
Wyse RE, Zamski E, Tomos AD (1986) Turgor regulation of sucrose transport in sugar beet taproot tissue. Plant Physiol 81: 478–481
Yamaya T, Filner P (1981) Resistance to acetohydroxamate acquired by slow adaptive increases in urease in cultured tobacco cells. Plant Physiol 67: 1133–1140
Yancey PH, Clark ME, Hand SC, Bowlus RD, Somero GN (1982) Living with water stress: evolution of osmolyte systems. Science 217: 1214–1222
Yano S, Ogawa M, Yamada Y (1982) Plant formation from selected rice cells resistant to salt. In: Fujiwara A (ed) Plant tissue culture 1982. Maruzen, Tokyo, pp 495–496
Yeo AR (1981) Salt tolerance in the halophyte *Suaeda maritima* L. Dum: intracellular compartmentation of ions. J Exp Bot 32: 487–497
Yeo AR, Flowers TJ (1986) The physiology of salinity resistance in rice (*Oryza sativa* L.) and a pyramiding approach to breeding varieties for saline soils. Aust J Plant Physiol 13: 161–173

Yeo AR, Yeo ME, Flowers SA, Flowers TJ (1990) Screening of rice (*Oryza sativa* L.) genotypes for physiological characters contributing to salinity resistance, and their relationship to overall performance. Theor Appl Genet 79: 377–384

Zenk MH (1974) Haploids in physiological and biochemical research. In: Kasha KJ (ed) Haploids in higher plants – advances and potential. University Press, Guelph, pp 339–353

Chapter 6

The Agricultural Use of Native Plants on Problem Soils

J. W. O'Leary

6.1 Evolution of Domestic Species

The origin and evolution of our present crop plants has been the cause for much speculation and conjecture. Our crops are largely not the result of any clear plan of action: there probably were many false starts, failures, and dead ends during the domestication of those plants (Heiser 1981). Some of the questions related to crop domestication, such as "where" and "when", are more easily answered, however, than others, such as "how" and "why" (Farrington and Urry 1985). Determining where and when crops were domesticated is accomplished by analysis and interpretation of physical evidence. Explaining how and why they were domesticated requires knowledge of motives and other intangible aspects of the people involved in the process of domestication. Thus, even though we may not agree on exactly how domestication occurred, there is general agreement that the process took place independently in several widely separated areas during the past 10000 years or so (Streuver 1971; Harlan 1975; Farrington and Urry 1985).

Based on proposed locales for the origins of agriculture, we can further speculate that this took place in favourable conditions. Fresh water and arable land were apparently in relatively great abundance during the domestication of our present crops. As a result, ability to tolerate adverse conditions (e.g. high salinity or high levels of specific toxic heavy metals or other minerals) were not criteria for selection during domestication. Therefore, cultivars of most of our present crops do not exhibit the necessary tolerance to permit their cultivation on many problem soils.

6.2 Limits to Improving Existing Crops

6.2.1 Genetic Variability Within Crop Species

Unfortunately, there may not be sufficient genetic variability for those traits related to stress tolerance within present crops to offer much promise of increasing tolerance through conventional breeding programmes (Bernstein

Monographs on Theoretical and Applied Genetics, Vol. 21
Ed. by A. R. Yeo and T. J. Flowers

1963; Abel and McKenzie 1964; Tal 1985). Salt tolerance, for example, is a complex trait assumed to be controlled by several genes (Tal 1985; Meyer et al. 1990). The relative contributions of each of those genes to the final phenotype are unknown, and even though conscious elimination of alleles conferring tolerance may not have intentionally occurred, useful variability could easily have been "lost" during the course of selection for other traits. In that case, no matter how much genetic recombination occurs, it will be impossible to assemble the genetic combination required to tolerate high salinity.

However, with a genetically and physiologically complex trait, such as salt tolerance, we might expect that there are some recombinations of genes within a crop species' genome that will result in higher salt tolerance than others. Assembly of those combinations through controlled breeding should yield progeny with greater salinity tolerance, even though one or more of the specific alleles essential for the combination giving maximum tolerance is missing. Furthermore, due to differences in the amount of genetic change during domestication, we should expect those recombinations to be more successful in some crops than in others. This scenario is consistent with the experience so far in breeding for increased salt tolerance in crops, and it also accommodates the more optimistic view of some (Epstein et al. 1980) regarding the amount of unexploited genetic variability for salt tolerance within present crops.

6.2.2 Salt Tolerance

Rapid progress is being made in development of techniques for transfer of genetic information among widely different plant species, or even genera. Thus, it may be possible to transfer the requisite genetic information responsible for salt tolerance, for instance, into important crop plants from alien genotypes (Dewey 1960; Bernstein 1963; Mudie 1974). However, if this approach is to be successful in improving the salt tolerance of crops, it is imperative that we know what genetic information to transfer. Even though the answer to this question may not be readily available, some progress has been made in that direction already. The best example may be the work of Gorham and co-workers (Gorham et al. 1985; Shah et al. 1987; Gorham 1990a, b, c; Gorham et al. 1990). Following the suggestion made by Dewey (1960), they have explored the potential of wild wheatgrasses for improving the salt tolerance of the crop species of the Triticeae, such as wheat, barley and rye. The have focussed on inheritance of the K/Na discrimination characteristic within the tribe, and even though there is not yet any direct proof of a relationship between enhanced K/Na discrimination and salt tolerance in the broad sense (Gorham 1990c), the work may serve as the stimulus for further work in this area.

6.2.3 Mineral Deficiency

In contrast to the problem of excessive soil salinity, some soils are deficient in content of specific minerals that are essential for plant growth. Since in many cases it is either too expensive, or impractical, to add the deficient element, there is a need for plants that could more efficiently use that element. To take calcium as an example, hope for finding calcium efficiency is based on the knowledge that some plants have evolved in environments where the soil level of calcium was high (calcicoles), while others (calcifuges) have evolved in areas where the soil levels of calcium were low (Clarkson 1965). It was recognised that there was genotypic variability for Ca uptake, transport and distribution (Marschner 1986), and it was shown that these differences in calcium use efficiency were heritable (Li and Gabelman 1990). Similarly, transfer of more efficient use of phosphorus (Schettini et al. 1987) and potassium (Figdore et al. 1989) from exotic germplasm into crop species has been demonstrated. Thus, it seems reasonable to expect that modifying crop plants to enable them to overcome deficiencies of particular nutrient elements in the soil can be accomplished through a combination of conventional breeding and the recently developed genetic transfer techniques.

6.2.4 Heavy Metal Toxicity

The case of excessive amounts of certain cations in the soil is very different (see also Chap. 7). With mineral deficiency, improved performance probably involves more effective use of limited amounts of the element through better delivery and concentration of that element at the sites of utilisation in the plant. With excessive heavy metals the element in question must be excluded from the sites of metabolism or "detoxified" through chelation or a similar process (Antonovics et al. 1971). Whatever the exact mechanism of tolerance may be, tolerance to one metal usually does not result in tolerance to another. The notable exception is Zn and Ni tolerance which co-occur in some plants.

Metal tolerance is known to be an inherited characteristic, usually dominant, although a large number of genes may be involved (Peterson 1983). Tolerant plants have been classified as:

- excluders, either at the root surface, or in loading for transport to shoots (i.e. root accumulators);
- index plants, in which the leaf concentration of the element in question reflects the concentration in the soil; or
- accumulators, in which the levels in the leaves exceed the levels found in the soil.

Even though all three types could be used, presumably, for cultivation on a soil with a high level of some heavy metal, the ultimate use of the plant could restrict the use to only one of those types. For example, if the metal is toxic to animals at low concentrations, then accumulators, and even index plants, might not be appropriate for use as crops if the plant is to be used for animal feed or human food.

6.2.4.1 Extent of the Resource Base

Unfortunately, there is only a small genetic resource base from which to draw in order to increase the heavy metal tolerance of crop plants. There are only a few species that regularly occur on highly contaminated soils in North America, primarily perennial herbs (Antonovics et al. 1971). In fact, with the exception of serpentine soils, no heavy metal-contaminated sites in North America have a unique or distinctive plant community associated with them (Wickland 1990). The most extensive and diverse heavy metal plant community in the world may be the metallophytes of south central Africa, but very little work has been done on evaluation of those plants for beneficial uses (Brooks and Malaisse 1990). Study of such populations of plants should be profitable, however, since recent reports indicate that there is considerable intrapopulation variation for various traits. Often such variation is as large as that observed between populations (Chen et al. 1986, Garbutt 1986; Verkleij et al. 1985). Thus, it may be possible to increase heavy metal tolerance in some crops by transferring tolerance from wild metallophytes. Whatever the mechanism of tolerance may be, there may well be a significant metabolic cost associated with addition of that trait to a crop plant, however. This could limit use of the modified plant strictly to problem soil areas. For example, in a study of eight naturally occurring populations of *Agrostis capillaris*, the relative growth rate under near-optimal conditions in a glasshouse was negatively correlated with increasing tolerance to copper and lead (Wilson 1988).

6.2.4.2 Heavy Metal-Binding Proteins

In some animals and fungi, heavy metals are detoxified through binding by metallothioneins (MTs), which are low molecular weight, cystein-rich proteins (Kagi and Nordberg 1979). These molecules apparently are not found in plants but the same role seems to be fulfilled in at least some plants by small peptides called phytochelatins (Grill et al. 1985). These molecules, which are not primary gene products, may not be as effective as MTs, and they typically are produced in response to acute heavy metal stress rather than to chronic exposure to toxic levels (Steffens 1990). In fact, it has been suggested that because of the sporadic and restricted occurrence of toxic levels of heavy metals, there probably has been insufficient selection pressure for evolution of a heavy metal detoxification system in plants (Steffens 1990). In that case, phytochelatins are viewed as existing for another purpose, and are recruited serendipitously by excess heavy metals. A chimaeric human MT gene has successfully been introduced into both *Brassica napus* and *Nicotiana tabacum* cells (Misra and Gedamu 1990). The resulting transgenic seedlings were unaffected by Cd levels up to 0.1 mM, while root and shoot growth in the controls were severely inhibited at those levels.

6.3 Availability of Alternatives

For some soil stresses, an alternative to genetic improvement of present crops, either by conventional selection or the use of exotic genetic variability, is to look to the native flora for potential crop plants. In contrast to the case of heavy metal tolerance, where there is a very limited native flora, high soil salinity offers a large group of plants from which to choose: the halophyte flora of the world. Furthermore, a considerable amount of effort has recently been devoted to exploiting this valuable resource, and the results have been very encouraging.

I will devote the rest of this chapter to a discussion of the general approach that has been taken in attempting to domesticate halophytes for use as crop plants and the results of those efforts. In addition to summarising progress toward development of such biological measures for mitigation of the problems of salinity, this may serve as a guide for the domestication of native plants for use as crops on other problem soils.

6.4 Methods of Domestication

A plan of action is required once a decision has been made to select candidates from a pool of "wild" plants for development as crop plants. In the case of halophytes, such a plan for accelerated domestication has been described (O'Leary 1984). In short, such a plan involves assessing the extent of the resource base, defining specific crop-use goals, screening as many candidate species/ecotypes as possible for their promise in meeting those goals, and determining the productivity of the most promising candidates under simulated agricultural conditions.

6.4.1 Extent of the Halophyte Resource Base

The genetic resource base of halophytes is extensive, encompassing over 1560 species in at least 117 diverse families (Aronson 1985, 1989). These species occupy the complete range of terrestrial environments from coastal marshes to arid deserts (Flowers et al. 1986). With 312 species, the Chenopodiaceae family represents about 20% of the total halophytic species (Flowers et al. 1986). Some halophytes have been used to a limited extent by native peoples for various purposes, including food, fibre and medicinal uses (Aronson 1989), but by and large, the greatest use has been as a forage for animals (Malcolm 1969; Le Houerou 1986; O'Leary 1988).

6.4.2 Use of Halophytes to Improve Rangeland Productivity

Collections of halophytes from around the world have been established for use in selecting the best candidates for extensive re-vegetation of rangelands (Malcolm et al. 1984). Thousands of hectares of rangeland have been planted with native and exotic halophytes to increase the productivity of those lands (Le Houerou 1986). Since most of those areas are arid or semi-arid, and the plants are dependent on rainfall for growth, the productivity is low compared to conventional agricultural crops, 5 tonne ha^{-1} yr^{-1} (dry weight) or less in most cases (O'Leary 1986). The important point, however, is that thousands of hectares of otherwise under-utilised land have been made productive, and support grazing of numerous animals. The majority of this activity has occurred in Northern Africa and the Mediterranean Basin (Le Houerou 1986) as well as in Australia (Malcolm 1969).

In some cases, re-vegetation with halophytes has primarily involved replacing or supplementing existing vegetation with faster-growing and/or more palatable species. Various members of the Chenopodiaceae, especially *Atriplex* species, have been used most often. The nutritional value of *Atriplex* has long been recognised (Jones 1970), and since this genus is represented in the flora of so many arid and semi-arid areas of the world, there is a long history of utilisation of *Atriplex* as supplemental feed by grazing animals (O'Leary 1986). These factors, coupled with the successful acclimation and adaptation of so many *Atriplex* species, such as *A. nummularia* and *A. canescens*, when introduced into new areas, are responsible for extensive use of these and other halophytes to increase productivity of rainfed rangelands around the world. In Australia and the United States, *Atriplex* species were planted on saline rangelands beginning in the 1900s (Jones 1970; Malcolm 1972). Such practices are now widespread throughout the arid and semi-arid areas of the world (McKell 1989), and some of those re-vegetation projects are immense. For example, in Libya alone, during the years from 1976 to 1979, 40000 hectares of land were planted with fodder shrubs, most of which were halophytes (Dumancic and Le Houerou 1981).

Grasses have been used as well for this purpose, primarily *Distichlis* (saltgrass) and *Leptochloa* (Kallar grass). In one case, 20000 ha of salt flats outside of Mexico City has been planted with *Distichlis spicata*, which is used for cattle forage. This may represent the largest area devoted to a single introduced halophyte species (BOSTID 1990). *Leptochloa fusca* has been planted extensively in Pakistan for the same purpose, particularly where soils are not only saline but also waterlogged (Malik et al. 1986; BOSTID 1990). In addition to high tolerance of waterlogged soils, other features of this plant which make it an attractive candidate for reclamation of problem soils are the relatively high K/Na ratio (for halophytes) in the leaves (Sandhu et al. 1981) and nitrogen fixation by free-living bacteria associated with the roots which can supply more than half of the nitrogen requirement of *Leptochloa* (Malik et al. 1986). *Distichlis* also is tolerant of flooded soils, but in contrast to *Leptochloa*, it also is tolerant of drought. *Distichlis* also tolerates much higher salinity than *Leptochloa*. Both species are perennial, so once established, they persist for years, which reduces the overall cost of production.

6.4.3 Use of Halophytes as Irrigated Crop Plants

More recent interest has focussed on using halophytes as irrigated crops in problem soil areas. Not only would irrigation increase productivity considerably, but if suitable halophytic crops can be identified, the potential applications become even greater. Use would not be limited to areas with problem soils due to high levels of salinity. For example, there are vast areas where crops cannot be grown because the water supply is inadequate. In many of those areas, however, there are abundant supplies of saline water, either underground or offshore. Even though the soils at these sites may not be considered problematic in their unirrigated state, once highly saline water is applied, the soil solution to which the plant roots are exposed is comparable to that on saline soils to which fresh water has been applied. In fact, irrigation of non-saline soils with saline water probably is far less problematic than irrigation of formerly agricultural soils that have become salinised, even if fresh water is used. One has to wonder what caused the salinisation of the soils in the first place, and the most likely answer is drainage problems. If the drainage problems are due to natural impediments rather than poor management, the cost of solving the problems may be prohibitive, and availability of crops that are tolerant of high salinity will not eliminate the need for adequate drainage. It is important to keep in mind that good drainage is required when using saline water for irrigation.

6.4.3.1 Use as Forage Crops: Productivity

Many halophytes have been screened for their potential use as forage plants. The dry matter productivity of these species, even when irrigated with highly

Table 6.1. Productivity of irrigated halophytes (kg DW/h/y)

Species	Yield	Reference
Atriplex lentiformis	17900	O'Leary et al. (1985)
Atriplex canescens subsp. *linearis*	17200	O'Leary et al. (1985)
Atriplex barclayana	8600	O'Leary et al. (1985)
Atriplex nummularia	8000	O'Leary et al. (1985)
Atriplex nummularia	10000–15000	Le Houerou (1986)
Atriplex nummularia	15300	Pasternak et al. (1985)
Salicornia bigelovii	15400	O'Leary et al. (1985)
Salicornia bigelovii	28000	Glenn et al. (1991)
Batis maritima	17400	O'Leary et al. (1985)
Spartina patens	14400	Gallagher (1985)
Distichlis palmeri	13600	O'Leary et al. (1985)

saline water, can approximate that of glycophytic conventional forage crops irrigated with fresh water (O'Leary 1988). Most of the promising candidates are chenopods, but a few grasses also are potentially useful (Table 6.1). It should be stressed that the yields in Table 6.1 are not corrected for ash content, which can be appreciable. Even when grown under non-saline conditions, the ash content of halophyte foliage can be 10 to 20% of the dry weight, and when grown on salinities equivalent to seawater, the ash content can be 40 to 50% (Glenn and O'Leary 1984). Thus, the reported yields of many halophytes must be adjusted downward. Nevertheless, some of the yields still are appreciable enough to be considered within the range of "good" agricultural yields for forage, and certainly much greater than the yields of these species on rainfed rangelands (O'Leary 1986; Watson et al. 1987).

Nutritional Value. In general, a high content of crude protein, nitrogen-free extract, calcium and phosphorus, combined with a low fibre content indicate high feeding value for ruminant monogastric animals, and *Atriplex* scores reasonably well in this regard (O'Leary 1988; Watson 1990). In conventional forage such as alfalfa, crude protein ranges from 15 to 30% crude protein (Hartman et al. 1981). By comparison, *Atriplex* species have been reported to have crude protein ranging from 13 to 20%, at least in leaf material (O'Leary et al. 1985). Since the crude protein values typically represent calculated values based on the assumption that all of the Kjeldahl-N is protein nitrogen, and halophytes typically contain substantial levels of other non-protein nitrogen such as proline and quaternary ammonium compounds, the reported values for halophytes probably overestimate true protein content (O'Leary et al. 1985). Another consideration is that most of those reported values are from experimental plants where the sample collection procedure resulted in primarily, if not completely, leaf material whereas if the plants are grown and harvested using conventional farm practices and equipment, the sample of cut and baled forage will consist of both leaves and stems. The percentage of the harvested biomass represented by stems can approximate 50% (Watson et al. 1987), and the resultant crude protein levels of cut and baled forage from several *Atriplex* species grown on a commercial farm and mechanically harvested reflect that composition, being in the range of 7 to 13% for seven species and three harvests (Watson and O'Leary 1993). Fibre content is good compared with alfalfa (Watson 1990). In fact, in a comparison of eight different families of forages, the Chenopodiaceae had the lowest fibre content (Larin 1947).

Palatability. Successful use of halophytes as a forage by animals, whether as browse on a range or as a component in a feed mix, depends on acceptability by the animal. Many plants, particularly "wild" ones, produce chemicals in their leaves that reduce their palatability. That many of these chemicals increase in response to herbivore pressure strongly supports the hypothesis that there is great survival value associated with their presence (Chew and Rodman 1979). Other evidence that the presence of the chemicals is a specific response to herbivory is based on the observation that the plants apparently perceive the difference between being clipped with a blade and being clipped by a grasshopper or cow (Janzen 1979). The saliva presumably makes the difference. Many of the compounds are alkaloids, and they are concentrated in those parts of the plant where herbivore attack would have the greatest negative impact (McKey 1974). For example, saponins are commonly found in high levels in chenopod leaves (Cheeke 1976) and seeds (Coxworth and Salmon 1972). Levels of these compounds can be quite variable, even within members of the same species. McKell (1989) describes a study in which 700 individual shrubs of *Atriplex canescens* were classified as to the degree of grazing intensity they experienced when contained within enclosures with grazing animals. One-third of the shrubs were grazed intensely, one-third moderately, and one-third were only lightly grazed. Analysis of the leaves indicated that saponin content was about one-third higher in the least palatable plants than in the ones most intensely grazed. In addition to the anti-feedant activity of saponin, there is a growth inhibition associated with it when ingested by animals (Cheeke 1976). It is possible to remove saponins by extraction in hot water or ethanol (Coxworth et al. 1969). It also has been possible to counter the growth-inhibiting effects of saponins by adding sterols to the feed (Coxworth and Salmon 1972), especially cholesterol (Cheeke 1976; Glenn et al. 1991).

Oxalic acid has been found in high levels in some members of the Chenopodiaceae (Goodin 1979), but the amounts found in the leaves of those species of *Atriplex* that have been considered good candidates for use as cultivated forage plants is not high enough to be a problem (Ellern et al. 1974; Goodin 1979). Cyanolipids in seeds can be a problem also (Siegler 1979). For example, the seed oil of *Ungnadia speciosa* keeps well and has physical properties that make it an attractive vegetable oil, but the presence of cyanolipids renders it useless because it is so toxic.

The high salt content also reduces the utility of the halophyte foliage as animal feed. The amount of halophyte forage that the animal will voluntarily eat is a function of the salt content, and most animals will only eat enough to provide the daily maximum tolerable salt intake, after which they just stop eating. One way to overcome this problem is to mix the halophyte forage with other feed materials (O'Leary 1986). In fact, it seems that the combination of the high ash content, along with the relatively low crude protein and energy contents of the foliage, dictates that it be blended or supplemented with other feed material as a practical measure (Watson and O'Leary 1993).

6.4.3.2 Potential for Use as Direct Food Crops

The most likely way in which halophytes could serve as direct food crops for humans is as seed crops. Fortunately, high ash content is not a problem with seeds of halophytes (O'Leary et al. 1985). The selective discrimination against sodium and chloride during phloem loading in source leaves results in low levels of these elements arriving in the developing fruits/seeds in halophytes just as in glycophytes (Hocking and Pate 1977; Hocking 1982). As a result, not only the levels of these two elements, but also the total salt content of the seeds and fruits are low (O'Leary et al. 1985). Of course, if the seeds are to be valuable as food sources, their nutritional value must be reasonably high, and the seed yield must be high enough to make it economically feasible to grow the plants as seed crops. Three halophyte genera that have been investigated for their potential as human food sources are *Distichlis*, *Kosteletzkya*, and *Salicornia* (Table 6.2).

Distichlis palmeri. *Distichlis* produces a seed that resembles wheat both in size and in composition. It is very high in starch and presumably could serve similar roles as wheat in foods. Preliminary results of milling, baking and taste tests indicated that products baked from *Distichlis* flour compared favourably with products made from wheat flour (Yensen and Bojorquez de Yensen 1987). The fibre content is about three times that of wheat, and the bran content reportedly is 40% by weight. There are no agricultural yield data available yet for this plant. *Distichlis* is dioecious, which means that only approximately half of the plants in a field (the females) will be seed-bearing. On the other hand, it is a perennial, which might be an advantage if, once established, the stand will yield a seed harvest for several years.

Kosteletzkya virginica. *Kosteletzkeya* has a moderate protein content and a modest fat content, but it has the highest fibre content of the three. Linoleic acid comprises about 55% of the lipid fraction of the seed, which makes it an attractive potential source of vegetable oil for human use (Islam et al. 1982). The seeds do not contain gossypol, as do the seeds of other members of the Malvaceae, such as cotton and okra, but the oil does contain cyclopropene fatty acids. The maximum salinity tolerated is about 2.5%, and the highest seed yield reported is 1400 kg/ha (Gallagher 1985).

Salicornia bigelovii. The halophytic seed crop that has been most intensively investigated to date is *Salicornia*. It not only has a high protein and oil content in the seed, but the linoleic acid level in the oil is about 75% (Glenn et al. 1991),

Table 6.2. Nutritional content of some halophyte seeds (%DW)

	Distichlis palmeri[a]	*Kosteletzkya virginica*[b]	*Salicornia bigelovii*[c]
Protein	8.7	24.9	31.2
Carbohydrate	79.5	–	–
Fat	1.8	15.4	28.2
Fiber	8.4	28.1	5.3
Ash	1.6	4.2	5.5

[a] Yensen and Weber (1986).
[b] Gallagher (1985).
[c] Glenn et al. (1991).

ranking it near the top of the list of all vegetable oils. This high content of unsaturated fatty acids makes it attractive as a vegetable oil for human consumption, but it also makes it highly vulnerable to rancidity, which could be a serious problem in those arid or semi-arid areas where the crop is most likely to be grown unless adequate storage conditions are provided. The most attractive feature of *Salicornia* may be the ability of the seeds to germinate under full-strength seawater, in contrast to many halophytes, which require fresh water for germination. Several years of field trials in northern Mexico have been conducted, and the 6-year mean seed yield was 2000 kg/ha (Glenn et al. 1991). Furthermore, it has been demonstrated that the seeds can be planted, harvested, and processed with conventional farm and processing equipment using standard methods. The resultant seed meal, after oil extraction, is almost 50% protein, which makes it an attractive animal feed supplement. The meal does contain saponin, which is an anti-feedant for certain animals, but if it can be removed economically, the saponin could be one more valuable by-product.

6.5 Time Scale for Agricultural Development

How long might it take to domesticate a wild halophyte, i.e. to develop a crop plant from its wild progenitor? The United States Department of Agriculture has proposed an optimistic timetable for such a process, in which seven stages are proposed for evaluating, developing and commercialising a new crop, starting with a wild species (Knowles et al. 1984). The overall process is projected to take at least 10 years. The early stages of germplasm collection and evaluation, utilisation studies and agronomic evaluation can probably be accomplished within about 7 years, but time required for breeding for improvement, production and processing scale-up, and commercialisation phases are less certain. Once started, breeding to improve the crop could continue indefinitely, depending on the type of crop. Even the evaluation process is dependent on the type of crop that is being evaluated. For example, if the potential use for the crop is as a forage, the entire process is less complex than if the crop is intended to be used as a raw material for a finished product. In the latter case, the finished product needs to be evaluated as well as the crop plant per se. This may require involvement of industrial partners, in addition to the agricultural researchers. Examples of such crops are kenaf, where the finished product is paper, and jojoba, where the finished product is a cosmetic.

Among the potential halophyte crops are *Atriplex* species intended to be used for forage and *Salicornia* and *Distichlis*, intended to be used as sources for vegetable oil or flour, respectively. *Atriplex* has already been evaluated for its potential as a forage: it already has been grown on commercial farms where it was irrigated with saline water and harvested with conventional farm machinery, and it is being used as a component in animal feed mix (Watson 1990; Watson and O'Leary 1993). There probably is not much improvement that is required to make wide scale use of *Atriplex* as a saline water-irrigated forage crop more likely, although it may be worth increasing the protein content of the foliage. The present protein level is not preventing the adoption of *Atriplex* or other halophytes as irrigated crops, however. Additional "fine-tuning" of the agronomic requirements of this crop is needed.

The most pressing need is to address the soil/water management practices. This issue must be resolved before any halophytic irrigated crop becomes a reality. The Scylla-Charybdis analogy used so effectively by Greenway and Munns (1980) to describe the dilemma faced by plants growing in saline environments could also be used here to describe the dilemma faced by the halophyte farmer. It would be desirable when using highly saline irrigation water to have soils that were very permeable, to insure adequate leaching. That is, to prevent salt accumulation in the soil, it would be important to apply enough irrigation water and have high enough soil porosity to flush the soil thoroughly and prevent the salinity of the soil solution from exceeding the salinity of the irrigation water, which is already high. However, such soils will have low water-holding capacity, so irrigation frequency will have to be high, perhaps even daily in a hot, dry environment, especially when the crop has reached the stage where the Leaf Area Index is high (Glenn and O'Leary 1984). Also, soils that do not retain ions such as sodium and chloride will not retain nitrates, phosphates and other desirable ions. Therefore, fertilisers may have to be applied in low doses at high frequency, maybe even with every irrigation. High irrigation inputs also require effective deep drainage otherwise the (saline) water table will rise. These problems of high volume/high frequency irrigation and frequent fertilisation would be mitigated considerably if the soil had higher water-holding capacity, but the problem of salt accumulation becomes more likely with such a soil. Thus, should the halophyte farmer strive for the soil high in sand and low in clay or organic matter content, or vice versa? Where along this continuum is the satisfactory range, and how extensive is it? These questions still need considerable attention before halophyte farming becomes a reality, but with a reasonable effort, the required answers could be obtained in 3 to 5 years, time enough so that it would not be unrealistic to expect to see forages obtained from halophytic crops irrigated with highly saline water by the turn of the century.

For the seed crops, where a finished product is involved, there are additional steps required, including processing and product preparation and evaluation. In addition, since the harvested part of the crop is seed rather than the entire above-ground biomass, there are other attributes of the crop that require attention, and probably improvement through breeding. This may include such things as photoperiodic control of flowering, lodging, premature shedding of the seed, and uneven ripening of the seed. It is quite common to find that the characteristics of seed production and dispersal that make a plant a successful wild plant make it an undesirable crop plant. How long does it take to make these kinds of improvements? A good model in this context is triticale. Even though it was the product of hybridisation between two domesticated plants, wheat and rye, the hybrid had many crop characteristics that needed improvement. In fact, 25 years ago triticale plants were generally tall (lodging-susceptible), late maturing, photoperiod-sensitive, and partially sterile with shrivelled seed (Villareal et al. 1990). Those problems, except for the shrivelled seed, have been corrected during the past quarter century through breeding, and today more than a million heactares in some 30 countries are planted with triticale. During the past 20 years, the seed yield has doubled (Villareal et al. 1990), also the result of breeding. This plant has progressed in that short time from essentially a wild plant used entirely for animal feed to one of the most promising human food sources today. The starch content is the same as wheat and rye, protein content is slightly higher, amino acid balance is better than wheat, and the lysine content is 12% higher than wheat. This is the result of a large, intensive, world-wide effort directed toward improvement of the crop characteristics of this "wild" plant. These results encourage an optimistic outlook for the prospect of making similar improvements with *Salicornia*, *Distichlis* and other halophytes.

6.6 Conclusions

What are considered problem soils today, in the context of contemporary agriculture, will inevitably have to be used for cultivation of crops. The problem of adapting crops to grow successfully on such soils is more complex and less amenable to solution in some cases than in others. The most difficult case may be development of crops tolerant to high soil salinity. It is widely recognised that salt tolerance is probably a complex trait requiring harmonious interplay of several genes, sôme of which may have been lost from the genome of some crop species as a result of selection during domestication. If the required genetic variation no longer exists within a species, it may be possible to transfer the requisite genetic information from other genotypes or species. The development of gene transfer techniques that allow crossing breeding barriers between plants previously believed to be impenetrable is cause for optimism in this regard. On the other hand, successful use of such technology is dependent on the ability to define accurately which traits should be transferred, and the genes that control them. Much work remains to be done to realise that goal, but it is an attainable goal. The required recipe for tolerance to high levels of salinity is present in a wide array of plants, collectively known as halophytes, so intensive study of those plants may yield the needed understanding.

An alternative to transferring salt tolerance characteristics from halophytes into crop plants is to do the reverse, transfer crop characteristics into halophytes. That is, select halophytes, that already contain the required complex set of characteristics for salt tolerance, for use as crop plants, and through selection and breeding, improve the agronomic characteristics of those plants. Increase seed size and/or number, improve seed quality, eliminate seed shattering and uneven ripening, improve the stature of the plants to reduce lodging tendency and increase ease of harvesting, and otherwise make the plant more useful as a crop plant. In contrast to increasing salt tolerance, changing these characteristics is just the sort of thing that we presumably know how to do well. The experience with plants such as triticale suggest that achievement of these specific goals can be accomplished in a reasonable amount of time, once an intensive effort is directed toward that end.

On the other hand, successful use of halophytic crops on saline soils or under irrigation with saline water is dependent on adequately addressing problems associated with soil/water management. The previously assumed biological barrier to crop growth under highly saline conditions in the soil has been shown to be surmountable. Plants can be grown on highly saline soils and have high productivity, and they should be amenable to improvement to increase their attractiveness as crop plants. If the cultural practices under those conditions can be successfully defined, then it may be possible after all to make effective use of much of the extensive area of those problem soils throughout the world.

Acknowledgements. I am especially grateful to Dr. Steven E. Smith for his helpful comments and suggestions in the preparation of this chapter.

References

Abel GH, McKenzie AJ (1964) Salt tolerance of soybean varieties (*Glycine max* L. Merrill) during germination and later growth. Crop Sci 4: 157–161

Antonovics J, Bradshaw AD, Turner RG (1971) Heavy metal tolerance in plants. Adv Ecol Res 7: 1–85

Aronson J (1985) Economic halophytes – a global review. In: Wickens GE, Goodin JR, Field DV (eds) Plants for arid lands. Allen and Unwin, London, pp 177–188

Aronson J (1989) Haloph – a data base of salt-tolerant plants of the world. Office of Arid Land Studies, University of Arizona, Tucson

Bernstein L (1963) Salt tolerance of plants and the potential use of saline waters for irrigation. Desalination Research Conference. National Academy of Sciences-National Research Council Publ 942, Washington, pp 273–283

BOSTID (Board on Science and Technology for International Development) (1990) Saline agriculture. Salt-tolerant plants for developing countries. National Academy Press, Washington

Brooks RR, Malaisse F (1990) Metal-enriched sites of South Central Africa. In: Shaw AJ (ed) Heavy metal tolerance in plants: Evolutionary aspects. CRC, Boca Raton, pp 53–73

Cheeke PR (1976) Nutritional and physiological properties of saponins. Nutr Rep Int 13: 315–323

Chen Z-Y, Scagel RK, Maze J (1986) A study of morphological variation in *Pseudotsuga menziesii* in southwestern British Columbia. Can J Bot 64: 1654–1663

Chew FS, Rodman JE (1979) Plant resources for chemical defense. In: Rosenthal GA, Janzen DH (eds) Herbivores – their interaction with secondary metabolites. Academic Press, New York, pp 271–307

Clarkson DT (1965) Calcium uptake by calcicole and calcifuge species in the genus Agrostis. Ecology 53: 427–435

Coxworth ECM, Salmon RE (1972) *Kochia* seed as a component of the diet of turkey poults: effects of different methods of saponin removal or inactivation. Can J Anim Sci 52: 721–729
Coxworth ECM, Bell JM, Ashford R (1969) Preliminary evaluation of Russian Thistle, *Kochia* and garden *Atriplex* as potential high protein seed crops for semiarid areas. Can J Plant Sci 49: 427–434
Dewey DR (1960) Salt tolerance of twenty-five strains of *Agropyron*. Agron J 52: 631–635
Dumancic D, Le Houerou HN (1981) *Acacia cyanaphylla* Lindl. as supplementary feed for small stock in Libya. J Arid Environ 4: 161–167
Ellern SJ, Samish YB, Lachover D (1974) Salt and oxalic acid content of leaves of the saltbush *Atriplex halimus* in the northern Negev. J Range Manage 27: 267–271
Epstein E, Norlyn JD, Rush DW, Kingsbury RW, Kelly DB, Cunningham GA, Wrona AF (1980) Saline culture of crops: a genetic approach. Science 210: 399–404
Evans LT (1980) The natural history of crop yield. Am Sci 68: 388–397
Farrington IS, Urry J (1985) Food and the early history of civilization. J Ethnobiol 5: 143–157
Figdore SS, Gabelman WH, Gerloff GC (1989) Inheritance of potassium efficiency, sodium substitution capacity, and sodium accumulation in tomatoes grown under low-potassium stress. J Am Soc Hortic Sci 114: 322–327
Flowers TJ, Hajibagheri MA, Clipson NJW (1986) Halophytes. Q Rev Biol 61: 313–337
Gallagher JL (1985) Halophytic crops for cultivation at seawater salinity. Plant Soil 89: 323–326
Garbutt K (1986) Genetic differentiation in leaf and whole plant photosynthetic capacity and unit leaf rate among clones of *Phlox paniculata*. Am J Bot 73: 1364–1371
Glenn EP, O'Leary JW (1984) Relationship between salt accumulation and water content of dicotyledonous halophytes. Plant Cell Environ 7: 253–261
Glenn EP, O'Leary JW, Watson MC, Thompson TL, Kuehl RO (1991) *Salicornia bigelovii* Torr.: an oilseed halophyte for seawater irrigation. Science 251: 1065–1067
Goodin JR (1979) Atriplex as a forage crop for arid lands. In: Ritchie GA (ed) New agricultural crops. Westview, Boulder, pp 133–148
Gorham J (1990a) Salt tolerance in the Triticeae: ion discrimination in rye and triticale. J Exp Bot 41: 609–614
Gorham J (1990b) Salt tolerance in the Triticeae: K/Na discrimination in *Aegilops* species. J Exp Bot 41: 615–621
Gorham J (1990c) Salt tolerance in the Triticeae: K/Na discrimination in synthetic hexaploid wheats. J Exp Bot 41: 623–627
Gorham J, McDonnell E, Budrewicz E, Wyn Jones RG (1985) Salt tolerance in the Triticeae: growth and solute accumulation in leaves of *Thinopyrum bessarabicum*. J Exp Bot 36: 1021–1031
Gorham J, Bristol A, Young EM, Wyn Jones RG, Kashour G (1990) Salt tolerance in the Triticeae: K/Na discrimination in barley. J Exp Bot 41: 1095–1101
Greenway H, Munns R (1980) Mechanisms of salt tolerance in non-halophytes. Annu Rev Plant Physiol 31: 149–190
Grill E, Winnacker E-L, Zenk MH (1985) Phytochelatins: the principal heavy-metal complexing peptides of higher plants. Science 230: 674–676
Hagemeyer J, Waisel Y (1988) Excretion of ions (Cd^{2+}, Li^+, Na^+, and Cl^-) by *Tamarix aphylla*. Physiol Plant 73: 541–546
Harlan JR (1975) Crops and man. American Society of Agronomy, Madison, WI
Hartman HT, Flocker WJ, Kofranek AM (1981) Plant science. Growth, development, and utilization of cultivated plants. Prentice-Hall, Englewood Cliffs, NJ
Heiser CB Jr (1981) Seed to civilization. The story of food, 2nd edn. Freeman, San Francisco
Hocking PJ (1982) Salt and mineral nutrient levels in fruits of two strand species, *Cakile maritima* and *Arctotheca populifolia*, with special reference to the effect of salt on the germination of *Cakile*. Ann Bot 50: 335–343
Hocking PJ, Pate JS (1977) Mobilization of minerals to developing seeds of legumes. Ann Bot 41: 1259–1278

Islam MN, Wilson CA, Watkins TR (1982) Nutritional evaluation of seashore mallow seed, *Kosteletzkya virginica*. J Agric Food Chem 30: 1197–1198
Janzen DH (1979) New horizons in the biology of plant defenses. In: Rosenthal GA, Janzen DH (eds) Herbivores – their interaction with secondary metabolites. Academic Press, New York, pp 331–350
Jones R (ed) (1970) The biology of *Atriplex*. CSIRO, Canberra
Kagi JHR, Nordberg M (1979) Metallothioneins. Birkhäuser, Basel
Knowles PF, Lessman J, Bemis WP, Blase MG, Burns EE, Burrows WC, Copp JH, Creech RG, Fike WT, Garrett RE, Hill LD, Idso SB, Jolliff GD, Jones Q, Miller JF, Purcell JC, Robinson RG, Sampson RL, Sumner DR, Thompson AE, Voss RD, Wedin WF, Wolff IA (1984) Development of new crops: needs, procedures, strategies, and options. Council for Agricultural Science and Technology Rep 102, Ames, IA
Larin IV (1947) USSR. In: Anonymous (ed) The use and misuse of shrubs as fodder. Joint Publ 10, Imperial Agricultural Bureaux, Oxford, pp 129–156
Le Houerou HN (1986) Salt-tolerant plants of economic value in the Mediterranean Basin. Reclam Reveg Res 5: 319–341
Li Y, Gabelman WH (1990) Inheritance of calcium use efficiency in tomatoes grown under low-calcium stress. J Am Soc Hortic Sci 115: 835–838
Malcolm CV (1969) Use of halophytes for forage production on saline wasteland. J Aust Inst Agric Sci 35: 38–49
Malcolm CV (1972) Establishing shrubs in saline environments. Western Australian Department of Agriculture Technical Bull 14, South Perth
Malcolm CV, Clarke AJ, Swaan TC (1984) Plant collections for saltland revegetation and soil conservation. Western Australian Department of Agriculture Technical Bull 65, South Perth
Malik KA, Aslam Z, Naqvi M (1986) Kallar grass: a plant for saline land. Nuclear Institute for Agriculture and Botany, Faisalabad, Pakistan
Marschner H (1986) Mineral nutrition of higher plants. Academic Press, New York
McKell CM (ed) (1989) The biology and utilization of shrubs. Academic Press, New York
McKey D (1974) Adaptive patterns in alkaloid physiology. Am Nat 108: 305–320
Meyer G, Schmitt JM, Bohnert HJ (1990) Direct screening of a small genome: estimation of the magnitude of plant gene expression changes during adaptation to high salt. Mol Gen Genet 224: 347–356
Misra S, Gedamu L (1990) Heavy metal resistance in transgenic plants expressing a human metallothionein gene. In: Lamb CJ, Beachy RN (eds) Plant gene transfer. Wiley-Liss, New York, pp 257–265
Mudie PJ (1974) The potential economic uses of halophytes. In: Reimold RJ, Queen WH (eds) Ecology of halophytes. Academic Press, New York, pp 565–597
O'Leary JW (1984) The role of halophytes in irrigated agriculture. In: Staples RC, Toenniessen GH (eds) Salinity tolerance in plants: strategies for crop improvement. Wiley, New York, pp 285–300
O'Leary JW (1986) A critical analysis of the use of Atriplex species as crop plants for irrigation with highly saline water. In: Ahmad R, San Pietro A (eds) Prospects for biosaline research. University of Karachi, Karachi, pp 415–432
O'Leary JW (1988) Saline environments and halophytic crops. In: Whitehead EE, Hutchinson CF, Timmermann BN, Varady RG (eds) Arid lands today and tomorrow. Westview, Boulder, pp 773–789
O'Leary JW, Glenn EP, Watson MC (1985) Agricultural production of halophytes irrigated with seawater. Plant Soil 89: 311–321
Pasternak D, Danon A, Aronson JA, Benjamin RW (1985) Development of the seawater agriculture concept. Plant Soil 89: 337–348
Peterson PJ (1983) Adaptation to toxic metals. In: Robb DA, Pierpoint WS (eds) Metals and micronutrients: uptake and utilization by plants. Academic Press, New York, pp 51–69
Sandhu GR, Aslam Z, Sakim M, Sattar A, Qureshi RH, Ahmad N, Wyn Jones RG (1981) The effect of salinity on the yield and composition of *Diplackne fusca* (Kallar grass). Plant Cell Environ 4: 177–181

Schettini TM, Gabelman WH, Gerloff GC (1987) Incorporation of phosphorus efficiency from exotic germplasm into agriculturally adapted germplasm of common bean (*Phaseolus vulgaris* L.). In: Gabelman WH, Loughman BC (eds) Genetic aspects of plant mineral nutrition. Nijhoff, Dordrecht, pp 559–568

Shah SH, Gorham J, Forster BP, Wyn Jones RG (1987) Salt tolerance in the Triticeae: the contribution of the D genome to cation selectivity in hexaploid wheat. J Exp Bot 38: 254-269

Siegler DS (1979) Toxic seed lipids. In: Rosenthal GA, Janzen DH (eds) Herbivores – their interaction with secondary plant metabolites. Academic Press, New York, pp 449–470

Steffens JC (1990) Heavy metal stress and the phytochelatin response. In: Alscher RG, Cumming JR (eds) Stress responses in plants: adaptation and acclimation mechanisms. Wiley-Liss, New York, pp 377–394

Streuver S (1971) Prehistoric agriculture. Natural History Press, Garden City, New York

Tal M (1985) Genetics of salt tolerance in higher plants: theoretical and practical considerations. Plant Soil 89: 199–226

Verkleij JAC, Bast-Cramer WB, Levering H (1985) Effects of heavy metal stress on the genetic structure of populations of *Silene cucubalus*. In: Haeck J, Woldendorp JW (eds) Structure and functioning of plant populations. North-Holland, Amsterdam, pp 355–365

Villareal RL, Varughese G, Abdalla OS (1990) Advances in spring triticale breeding. Plant Breed Rev 8: 43–90

Watson MC (1990) *Atriplex* species as irrigated forage crops. Agric Ecosyst Environ 32: 107–118

Watson MC, O'Leary JW (1993) Performance of *Atriplex* species under irrigated conditions and mechanical harvests. Agric Ecosyst Environ 43: 255–266

Watson MC, O'Leary JW, Glenn EP (1987) Evaluation of *Atriplex lentiformis* (Torr.) S. Wats. and *Atriplex nummularia* Lindl. as irrigated forage crops. J Arid Environ 13: 293–303

Wickland DE (1990) Vegetation of heavy metal-contaminated soils in North America. In: Shaw AJ (ed) Heavy metal tolerance in plants: evolutionary aspects. CRC, Boca Raton, pp 39–51

Wilson JB (1988) The cost of heavy-metal tolerance: an example. Evolution 42: 408–413

Yensen NP, Bojorquez de Yensen S (1987) Development of a rare halophyte grain: prospects for reclamation of salt-ruined lands. J Wash Acad Sci 77: 209–214

Yensen SB, Weber CW (1986) Composition of *Distichlis palmeri* grain, a saltgrass. J Food Sci 51: 1089–1090

Chapter 7
Metal Toxicity

T. McNeilly

7.1 Introduction: Sources of Toxicity

Problems posed to plants by metal toxicity in the soils of the world are basically of two kinds. The first kind are of natural origin. These arise either as a consequence of the nature of the parent material from which a particular soil is derived, or from the processes of soil formation. Such events tend to lead to toxicities due to the products of soil mineral decomposition under acid conditions, predominantly to aluminium and/or manganese and iron. The soils characterised by aluminium and manganese toxicities comprise some 40% of the world's land area given over to arable farming (Clark 1982), and as such, potentially pose a major constraint to the world's agricultural production.

The second type of metal toxicity in soils is anthropogenic in origin, having been imposed on soils by pollution originating from the increasing growth of the industrial and domestic impacts of humans on their environment. Acid soil development, leading to aluminium toxicity as a result of sustained use of acid-forming fertilisers, has occurred mainly through the nitrification of ammonium ions. This is a potential problem in the agricultural exploitation of prairie soils in the United States of America (Carver et al. 1988) and in certain soils in Australia under clover pastures (Williams 1980). The most widespread anthropogenic source of metal toxicities is from the mining and smelting of the heavy metals; copper, lead and zinc. Whereas the area affected by natural sources of aluminium and manganese extends to many millions of hectares, the effects of anthropogenic contamination with Cu, Pb and Zn are much more localised, varying from a few square metres in the case of slag left from smelting in 1500 B.C., to several hundred hectares in the lower Swansea Valley in South Wales.

Soil toxicity due to heavy metals is not, however, confined to areas affected by human activity. A number of sites of natural outcroppings of ultramafic serpentine rocks, variously rich in combinations of chromium, cobalt, copper, manganese and nickel occur throughout the world and give rise to soils which are inhospitable to plant growth (Wild 1965; Kruckeberg 1984; Baker and Brooks 1989; Morrey et al. 1989).

Toxicity due to aluminium and manganese affects large areas of agricultural land, and hence can impose direct constraints on potential crop production

Monographs on Theoretical and Applied Genetics, Vol. 21
Ed. by A. R. Yeo and T. J. Flowers

unless appropriate soil amelioration treatments are applied. Soil toxicity problems due to the heavy metals affect only local communities and have little or no impact on general agricultural production – although increased use of treated sewage on an extensive scale does have the potential to raise soil heavy metal levels significantly. Therefore with heavy metals the need is to limit the deleterious effects of movements into the food chain of toxic amounts of these elements as dust blow from contaminated sites, or in surface and/or percolating drainage waters. In contrast, there is a vastly greater need to address the problems of aluminium and manganese phytotoxicities. Whilst liming can reduce the toxic effects of both aluminium and manganese, this is not universally possible because of the large areas affected and the non-availability of abundant supplies of lime. The need is, therefore, to find plant material which not only survives, but grows and yields well in the presence of these metals.

It is now well established that tolerance to heavy metals in soil has evolved. In some cases such soils support endemic, and rare taxa (Wild 1964; Wild and Bradshaw 1977; Brooks and Malaisse 1989), as well as metal-tolerant ecotypes of more widespread species (Wild and Bradshaw 1977). A similar situation occurs at many of the sites where the source of the metal is pollution, rather than natural occurrence. Here ecotypes of a small number of widespread species are commonly colonisers (Bradshaw and McNeilly 1981), although some sites support isolated populations of what appear to be relicts from previously more widespread distributions (Lefèbvre 1967, 1968; Ingrouille and Smirnoff 1986). Although less fully investigated, aluminium tolerance has also evolved (Rorison 1960). There are similarities in the modes of origin of heavy metal-tolerant material and the early aluminium-tolerant races of certain crop species: both were apparently the product of natural selection. Tolerance to aluminium and manganese has been, and is being, actively sought in crop species such as the cereals, legumes, sunflower, tomato and others (Foy 1974, 1983a) and manganese tolerance in a similar but more restricted range of species (Foy 1983b). By contrast, only populations of those species known to have evolved tolerance in the wild through natural selection have been sought and used as they are in the rehabilitation of heavy metal contaminated land. The great interest that has been shown in heavy metal tolerance in plants has been predominantly because it has provided an outstanding model system with which to examine the processes of natural selection in the wild.

Whilst a contrast has been drawn between plant breeding and evolution in the wild in the origin of aluminium and manganese tolerance in crop species and heavy metal tolerance in wild species, success in the two processes is based on the same three requirements. Firstly, there must be variability within the targeted species in the character or characters which are to be subjected to selection. Secondly, some means is needed whereby some individuals which are superior to the majority can be identified easily. Thirdly, to at least some degree, the characters favoured by selection must be under genetic control and there must be no deleterious genetic correlations with other characters. These three requirements will be examined in relation to metal tolerance. It is first necessary, however, to provide a brief account of the methods which have been used to provide quantitative measures of tolerance, whether of heavy metals or of Al and Mn.

7.2 The Measurement of Tolerance

In order to assess whether a character is present within a species and, if so, the extent of its variation, and to proceed from this to a selection programme, it is clearly necessary to have some means of identifying, and preferably quantifying in a consistent way, the differences between the individual phenotypes under consideration. Reviews by Wright (1976), Blum (1988) and Scott and Fisher (1989) provide discussions of the methods available for detecting tolerance and the reader is referred to these and to Baker and Walker (1989) for further details.

In essence, two basic procedures are available. Firstly, plants can be grown in soils which have metal concentrations sufficiently high to be toxic to plants which are not tolerant. Secondly, plants can be grown in solution cultures in which metal concentrations are adjusted to a point at which they are toxic to non-tolerant plants. In both types of experiment, tolerance can be assessed over a range of metal concentrations.

Other methods have been used on a limited scale to investigate the effects of the heavy metals. These include the use of metal concentrations at which epidermal cells are killed, in order to provide a measure of cell resistance to metal ions (Repp 1963; Gries 1966); dry weight production in water culture experiments (Broker 1963); and measurements of the rate of photosynthesis and general growth (Baumeister 1954; Baumeister and Burghardt 1956; Wachsmann 1961).

The primary effect of both aluminium and the heavy metals is, however, on root growth, which is progressively inhibited as the metal concentration increases. In the case of the heavy metals, at sub-lethal concentrations, the primary root apex is killed and lateral roots begin to grow, giving the root the so-called "hedgehogging" appearance. This contrasts with aluminium toxicity, where lateral roots are apparently more sensitive to aluminium than the primary roots and the root systems have a stubby appearance. No equivalent (the clear and rapid effects of aluminium (Foy 1983a) or heavy metals), can be seen on plant shoots. As a consequence, various aspects of the root growth of plants, whether grown in solution culture or in soil, have been used to quantify the toxicity of, or tolerance to, these metals. From a critical assessment of the relative value of 12 measures of root growth and two of shoot growth of 7- and 14-day-old sorghum seedlings, Furlani and Clark (1987) concluded that the longest adventitious root and the net seminal root lengths provided the most sensitive measures of plant response to aluminium. Clearly, only adventitious root growth is available for measurement in adult plants. This is very similar to the experience gained with heavy metals.

7.2.1 Solution Culture Methods

These can be divided into series methods and parallel methods. In the former, a measure of growth of the same root is made, first in non-toxic conditions and then in toxic concentrations of the appropriate metal(s).

The parallel method involves root growth of different replicate groups of standardised plant material in control and in metal-containing solution under identical conditions, over a standard length of time. This is clearly of value mainly for inbreeding species, where genotype replication can be guaranteed, or with material such as grasses, where a single individual can provide potentially unlimited numbers of genetically identical ramets as tillers (Bradshaw and McNeilly 1981). The parallel method can also be used for a relatively small number of herbaceous species, such as *Rumex acetosa* (Thompson 1987) *Mimulus guttatus* (Allen and Sheppard 1971) and *Silene maritima* (Baker 1978), in which cuttings from individual genotypes will root readily.

In both methods, root elongation rates in control and toxic conditions are obtained and, from the ratio of one to the other, an index of tolerance or relative tolerance is obtained, usually expressed as a percentage of control root length (Wilkins 1957; Jowett 1958), where

$$\text{tolerance index} = \frac{\text{mean root length in the presence of the metal} \times 100}{\text{mean root length in the absence of the metal}}.$$

A cautionary note about the use of such tolerance indices in genetic studies has been made by Humphreys and Nicholls (1984) based upon possible bias introduced from control root length measures. Their (Humphreys and Nicholls 1984) data suggest that control and treatment (i.e. metal-treated) root lengths may be under independent genetic control. The combination of these values into an index may, they suggest, lead to confusing or at worst misleading interpretations of the results of a crossing programme. They also draw attention to the observation that control root lengths of plants from metal-contaminated soils tend to be generally less than those of plants from normal soils, associated with the generally lower nutrient status of the former. These differences can lead to an upward bias in the tolerance indices obtained for plants from toxic sites. Leading from this disparity in soil nutrient status, the nature of the background solution can also lead to biased tolerance estimates based on tolerance indices. A further discussion of these problems can be found in Macnair (1989).

Because of variation in the performance of different ramets taken from the same individual, measures of tolerance based on tolerance indices or root length values frequently require the measurement of the roots of large numbers of ramets. In addition, where aluminium tolerance is being examined, the acidity of the growth medium has to be monitored and maintained on a daily basis to ensure that the aluminium is not precipitated. Where tolerance of adult plants is being examined, solutions have to be changed regularly in order to maintain metal levels against uptake by the plants. These procedures are somewhat time-consuming, and Polle et al. (1978) suggested a simpler and quicker method than those relying on root measures. Their procedure involves no measurements of root lengths, and is based upon the ability of haematoxylin staining to locate aluminium in plants (Hoffer and Carr 1923) and the data of McLean and Gilbert (1927), who showed that less aluminium accumulated in the cortex of roots of aluminium-tolerant than aluminium-sensitive plant species. Polle et al. (1978) recommended their method for large-scale screening of seedling materials, using simple visual assessment of haematoxylin staining of roots grown in a series of increasing aluminium concentrations in solution culture. They were able to distinguish differences in tolerance between the five wheat cultivars they examined and their ratings agreed with those from a previous solution culture experiment (Moore et al. 1976).

Interestingly, Moore et al. (1976) described a solution culture method for determining aluminium tolerance that required only simple assessment: the recovery of root growth in wheat seedlings after what the authors termed an "aluminium insult" – a 48-h exposure to a concentration which would completely inhibit cell division and hence root growth. Such re-growth could be determined after 24 h. Seedlings in which re-growth had not occurred after 72 h did not recover. A further solution culture based method used for screening large numbers of seedlings (15 000–30 000) for aluminium tolerance and which avoids measurements in large seedling samples was described by McNeilly (1982). The rooting response of a representative sample of non-tolerant material to increasing aluminium concentrations in solution culture was measured, and a concentration was determined which just inhibited rooting. Variable germplasm is then screened at that concentration and those individuals in which rooting occurs can be selected as showing tolerance that is greater than the original sample. Screening for tolerance to other metals or to salinity can be made using this technique: it was successfully used by Ingram (1987) to select for improved copper tolerance and by Ashraf et al. (1986) for salinity tolerance in a number of grass species.

A number of methods which employ root growth in solution cultures as a means of assessing metal tolerance involve pre-germination of seed or growing of seedlings in solution without metal for a number of days prior to their transfer to metal-containing solution. Recent information (D. A. Thurman pers. comm.) has shown that significant differences in estimates of cadmium tolerance are found when the same material is grown in this way and when grown from the outset in solution containing cadmium. Differences in relative tolerance values of between 10 and 45% were recorded. No information is available about the possibility of some form of genotype by environment interaction affecting such responses, but should they occur, they could have a marked effect on the tolerance rankings of materials tested under these differing experimental conditions.

In the case of manganese toxicity, in contrast to that of the metals considered previously, the main effects are initially seen in leaves and plant shoots (Reid 1976; Scott and Fisher 1989). The site of control of the response appears to vary with species, being apparently in the root of lucerne (Ouelette and Dessereaux

1958) but sited in the shoot of soybean (Heenan and Carter 1976). As with aluminium tolerance, solution culture techniques are available for testing for tolerance/susceptibility to manganese. Mean dry matter production and root weight ratios have been used to assess manganese toxicity in a number of tropical and temperate legumes (Andrew and Hegarty 1969).

7.2.2 Soil-Based Methods of Testing Tolerance

Soil-based methods for assessing tolerance involve comparison of plant growth in soils which contain the appropriate metal(s) either naturally, or as artificially added salts. The first published data demonstrating the contrast in response of tolerant populations from mining sites and control populations to copper toxicity involved the addition of copper carbonate to a garden loam soil (Prat 1934). The use of salt addition to non-toxic soils has, however, seen little subsequent use in the study of heavy metal tolerance, most subsequent work having used solution culture methods. Early experiments screening for tolerance to copper and zinc in a range of grass and forage species (Gartside and McNeilly 1974; Walley et al. 1974) used highly toxic mine spoils with additions of varying proportions of non-toxic loam, since there were no survivors on the unamended mine spoils.

Acid soils having pH values within the range at which aluminium and manganese become soluble have, on the other hand, been widely used for field, glasshouse and laboratory estimation of tolerances to these metals. Aluminium in these soils is frequently quantified as the aluminium saturation percentage, where

$$\text{saturation percentage} = \frac{\text{Al}}{\text{Al} + \text{Ca} + \text{Mg} + \text{K} + \text{Na}}.$$

Lime is then added to reduce aluminium saturation percentage to a level at which toxicity symptoms are removed, thus providing an appropriate control soil.

Plant growth on such contrasting soils can then be assessed for whatever character(s) are chosen as appropriate and quantitative measures of growth parameters can be used directly or as relative tolerance indices as described previously for solution culture experiments.

The distribution of aluminium and manganese toxicity symptoms in soils appears to be relatively homogeneous within a single field and this allows extensive field trials to be carried out for germplasm assessment. This is in marked contrast to the situation for soil salinisation, where marked heterogeneity in soil salinity distribution over relatively small distances precludes meaningful field experiments for the determination of salinity tolerance (Richards and Dennett 1980; Richards 1983; see also Chaps. 2 and 3). Problems of using soil for screening for aluminium and manganese tolerance can arise, however, mainly from lack of control of exposure to the metal concerned and the levels of other nutrients, particularly calcium and phosphorus, as well as variation in the pH. This arises because soil chemistry is in a continuous state of dynamic flux with regard to chemical speciation of metals and of oxidation and hydrolysis, processes which also affect metal bio-availability. Nonetheless, the use of soil as a medium for assessment of tolerance to aluminium and manganese has been widely and successfully used, both in field (Gourley 1987), and glasshouse (MaClean and Chiasson 1966) and in Petri dishes under laboratory conditions (Hill et al. 1989).

The methods outlined above provide simple means which allow distinctions to be drawn between plant phenotypes displaying different degrees of tolerance to heavy metals, to aluminium and manganese, and to boron. They thus form the basis for carrying out selection, one of the three requirements for successful evolutionary change through breeding.

It is possible that the use of solution culture methods for assessing tolerance to metals may not necessarily reflect field performance in toxic soil. A number of reports have examined the relationship between the aluminium tolerance rankings of lines/cultivars of a number of species grown in solution culture and in the field (Table 7.1). In general, there is an acceptable relationship between the two methods (Scott and Fisher 1989), but one which they consider acceptable in general terms only. However, since the rapid methods using solution

Table 7.1. Relationships between estimates of metal tolerance in short-duration laboratory-based experiments, and long-term field experiments, (1) aluminium tolerance in four crop species, and (2) copper tolerance in three grass species

Species	Comparison	r	Reference
1. Aluminium tolerance			
Barley	Root wt in solution Root wt in soil	0.75***	Reid (1976)
Rice	Relative root length Grain yield in field	0.64+	Howeler and Cadavid (1976)
Lucerne	Dry matter in solution Dry matter in soil	0.73**	Campbell et al. (1989)
Wheat	Relative root length in solution Relative grain yield in field	0.51***	Campbell and Lafever (1976)
2. Copper tolerance			
Agrostis capillaris		0.94***	
Dactylis glomerata	15-day-old seedlings Adult plants	0.83***	Ingram (1987)
Lolium perenne		0.85***	

** and *** denote correlation significant at $P < 0.01$ and 0.001 respectively.
+ Reported as highly significant.

culture provide a method for initial screening only, subsequent field trials would ensure that any false acceptances would rapidly become evident.

7.3 Variability in Wild Species

7.3.1 Tolerance to Heavy Metals

Whilst toxicity due to the presence of heavy metals, singly or in combination, is the over-riding selective agent on metal contaminated soils, many other factors, such as extremely low nutrient status (particularly nitrogen and phosphorus), low pH and poor water holding capacity, are also characteristic of many mine spoils. A considerable number of species grow on soils in the British Isles with similar features to mine wastes, but which lack their heavy metal toxicities. Despite this, only a relatively small number of species are known to colonise heavy-metal-contaminated soils. It appears that those species which are absent from toxic mine spoils, despite their frequent proximity to them (Table 7.2), are unable to colonise them because the plants lack the necessary genetic variation in tolerance to the appropriate metal.

This has been shown in experimental screening of large numbers of individuals from several grass and forage species to copper (Gartside and McNeilly 1974; Ingram 1987). The evidence from these early screening experiments showed that it was relatively straightforward to select individuals of *Agrostis capillaris* that were tolerant to copper (Gartside and McNeilly 1974)

Table 7.2. Species growing in the copper waste and in the non-toxic soil at Glasdir copper mine, N. Wales

Species growing on copper waste	
Agrostis vinealis	*Deschampsia flexuosa*
Agrostis capillaris	
Calluna vulgaris	
Species growing in non-toxic soil surrounding the mine site	
Agrostis vinealis	*Pinus sylvestris* (planted)
Agrostis capillaris	*Poa trivialis*
Betula pendula (planted)	*Pseudotsuga menziesii*
Calluna vulgaris	*Pteridium aquilinum*
Digitalis purpurea	*Ranunculus ficaria*
Galium saxatile	*Ranunculus repens*
Hedera helix	*Rhododendron ponticum*
Juncus effusus	*Rubus fruticosus*
Luzula campestris	*Taraxacum officinale*
Lysimachia nemorum	*Ulex europaeus*
Pinus contorta (planted)	

and zinc (Walley et al. 1974). However, it has subsequently been shown using seed material from a number of different populations, even from within *A. capillaris*, the most common coloniser of contaminated mine wastes, that occurrence of the appropriate genes for tolerance to cadmium, copper, lead, nickel and zinc is a feature of individual populations, and not of the species as a whole (Symeonidis et al. 1985). Further work has shown that this absence of appropriate variability in tolerance in particular populations can be a constraint on the evolution of zinc-tolerant populations of *A. capillaris* on the zinc-contaminated soils which occur beneath electricity pylons (Al-Hiyaly 1989). Thus, not only is the absence of appropriate variation a constraint on artificial selection, it can also be a constraint on selection in the wild.

An examination of the floras of natural metal-rich outcrops in the Great Dyke in Zimbabwe and those of adjacent non-metalliferous soils shows that the former represent only a small proportion of the latter (Wild and Bradshaw 1977). This would again seem to suggest, particularly in view of the very great age of these metal-rich outcrops (which have probably been in existence since the origins of the angiosperms during the Cretaceous) and the possibilities for colonisation that this implies, that only a limited number of species are able to evolve metal tolerance. The occurrence of different species complexes on similar sites within the same area may again reflect the random distribution of genes for tolerance across populations of the same species.

7.3.2 Tolerance to Aluminium and Manganese

Differences in the tolerance to aluminium of plant genera was shown by Rorison (1960) to be a significant factor in their distribution and he concluded that the absence of the calcifuge species *Scabiosa columbaria* from acidic habitats was due to aluminium toxicity, to which *S. columbaria* is susceptible. Rorison subsequently showed that *Deschampsia flexuosa*, a species characteristically found only on very acidic soils, is able to colonise them because of the high aluminium tolerance which it possesses (Rorison 1969). A close correlation between field distribution of 15 species and sensitivity of their seedlings to aluminium was shown by Grime and Hodgson (1969). They again reported the high resistance of *D. flexuosa*, and included in this category the calcifuges *Nardus stricta* and *Holcus lanatus*, whereas the calcicole species *Sanguisorba europea*, *Centaurea nigra* and *Leontodon hispidus* were very sensitive to aluminium. From previous arguments it would appear that the ecological amplitude of these species may be limited primarily by their lack of genetic variability in aluminium tolerance.

In addition to differences between genera, differences in reaction to aluminium are found, perhaps not unnaturally, between related species within a genus: this again relates to their ecology. At low pH, aluminium is more toxic to the calcicole *Carex lepidocarpa* than to *C. demissa*, which tends to behave as a calcifuge (Clymo 1962). In a similar way, four of the five British species within the genus *Agrostis* show a broad relationship between their ecological distributions in relation to soil pH and the degree of aluminium tolerance they show (Clarkson 1966). *A. setacea*, a species exclusively of sandy acid and peaty heaths and moors, was highly tolerant to aluminium; *A. canina* and *A. capillaris*, of somewhat less acid soils, had intermediate tolerance, whilst *A. stolonifera*, a species of more base-rich soils, was the least tolerant. Of these four species, *A. capillaris* has the greatest ecological amplitude, occurring on calcareous soils over limestone, on neutral soils and also on acid upland heaths. *A. stolonifera* occurs widely across the spectrum from base-rich to neutral soils.

Differential tolerance to aluminium and manganese has been demonstrated in populations of *Melilotus alba* (Ramarkrishnan 1968) and *Hypericum perforatum* (Ramarkrishnan 1969) from acidic and calcareous soils. Ramarkrishnan suggested that the contrasting distributions of the two population types was due to their differences in tolerance to aluminium and manganese. In a broader context, this would seem to suggest that more widespread species would show a diversity of tolerance to aluminium paralleling their ecological diversity. The same argument would seem, logically, to apply to other species of wide ecological amplitude such as *Anthoxanthum odoratum*, *Festuca ovina*, *F. rubra* and *Koeleria gracilis*. However, two more recent studies which have examined populations of *Chamaenerion angustifolium* (de Neeling and Ernst 1986) and *Succisa pratensis* (Pegtel 1986), both from acid and base-rich soils, report no differences between populations in response to aluminium. By contrast, both species showed population differences in response to manganese. The situation is clearly a complex one. In *Chamaenerion*, although the plants from acidic sites grew better at lower manganese levels, at the highest manganese concentrations, growth of the acidic population was severely restricted, whilst the calcareous population was unaffected. It was suggested (de Neeling and Ernst 1986) that this effect was due to the lower Mn/Fe ratio in the leaves of the alkaline plants. In *Succisa*, growth of the calcareous population was less inhibited than the acidic population and this was thought due to phenotypic plasticity in response to aluminium and manganese (Pegtel 1986).

It would seem unwise at this stage, from the somewhat limited information available about a very restricted number of species, to postulate too broadly about the underlying impacts of aluminium and manganese tolerances (or their absence) on species/ecotype distribution and whether this is determined, as it is with heavy metal tolerance, by lack of appropriate variability.

7.4 Variability in Cultivated Species

There is a considerable amount of information about the tolerance of crop species, particularly to aluminium and to manganese. By contrast, there is little information about crop tolerance to heavy metals. This reflects the differences in the extent of problems of aluminium and manganese toxicities in acid soils and those due to the heavy metals.

7.4.1 Tolerance to Zinc

The possibility that heavy metal levels in soil might rise as a consequence of increased use of domestic refuse as soil conditioners and as landfill has stimulated some assessment of the reaction of a small number of crop species to zinc, the main problem seen to be associated with domestic refuse dumping. The reactions of three crop species, snapbean (*Phaseolus vulgaris*), cucumber (*Cucumis sativa*) and maize (*Zea mays*) to added soil zinc was examined by Walsh et al. (1972), the amounts of zinc used reflecting estimated inputs from zinc-based fungicides. Species reactions were different, in that addition of 367 kg Zn ha reduced the yield of snapbean, but maize and cucumber yields were unaffected. Previous work with snapbean (Polson and Adams 1970) had shown that there were cultivar differences in response to increased zinc in the growing medium. In soybean, White et al. (1979) were able to distinguish tolerant, normal and susceptible classes based upon differences in plants grown in a glasshouse in soil amended with zinc sulphate.

7.4.2 Tolerance to Aluminium and Manganese

Different crop species are known to vary in their responses to aluminium and manganese: like wild species, they show a spectrum of tolerance ranging from the more aluminium tolerant, such as cowpea, *Vigna unguiculata*, groundnut, *Arachis hypogea*, and the potato, *Solanum tuberosum* (Little 1988), to the very sensitive alfalfa, *Medicago sativa* (Rechcigl et al. 1988). Within the temperate cereals there is a spectrum of aluminium tolerance which ranks rye > oats > hexaploid wheats > barley, with the inter-generic hybrid *Triticale* lying between its two parents in tolerance. In the past, the geography of exploitation of the four crops in the British Isles may have reflected these differences in their tolerances of acid soils.

Reviewing plant adaptation to acid soils, Foy (1988) lists 23 species for which there is evidence of intra-specific variation. These and some recent additions are listed in Table 7.3. It is interesting, however, to examine aluminium tolerance in three of these species, wheat, sorghum and rice, in relation to their ecological amplitude, particularly with respect to soil acidity.

7.4.2.1 Wheat

There are, as in wild species, a number of cultivated species which are grown across a wide spectrum of soils of differing acidity, which show related variation in aluminium toxicity. Probably the most important example is found in the

Table 7.3. Species in which variation in aluminium tolerance has been reported. (Based on Foy 1988 except where otherwise referenced)

Avena sativa	*Leucaena* spp.
Hordeum vulgare (Reid 1976)	*Amaranthus* spp.
Sorghum bicolor	*Glycine max*
Oryza sativa	*Lotus corniculatus*
Secale cereale	*Medicago sativa*
Zea mays	*Trifolium pratense*
	(Baligar et al. 1987)
X. tritosecale (Ittu and Saulescu 1988)	*Trifolium subterraneum* (Evans et al. 1987)
	Trifolium repens (Cardus 1987)
Saccharum officinarium	*Lespedeza cuneata*
Lolium spp.	*Lycopersicum esculentum*
Phalaris aquatica	*Nicotiana tabacum*
Phalaris arundinacea (Culvenor et al. 1986a)	*Manihot esculenta*
Phaseolus vulgaris	*Ipomoea batata*
Helianthus annus	
Solanum tuberosum (Little 1988)	
Vigna unguiculata (Horst 1987)	

hexaploid wheats. The conscious selection and breeding of aluminium-tolerant wheat cultivars began in Brazil in 1919 in Rio Grande del Sol State where soil aluminium toxicity was so great as to eliminate all non-adapted (i.e. non-Al-tolerant) material. By contrast, local land races or cultivars were already widely grown on these soils with no addition of lime and little if any additional nutrient inputs (da Silva 1976). These original tolerant land races/cultivars presumably had their origins from variable source materials introduced at some time since the Portuguese colonisation of Brazil. They were the product of unconscious selection, by local farmers, of the products of natural selection on these acid aluminium-toxic soils. The recognition that there was a correspondence between the deaths of non-adapted plants and the aluminium toxicity of the soils is attributed to Pavia in 1944 (da Silva 1976), and the first experimental confirmation that aluminium toxicity was responsible came from Araujo in 1953.

A parallel situation appears to have occurred in Kenya, where the cultivars K. Knogomi, K. Tembo and Romany have apparently never been consciously selected for tolerance to aluminium, but nonetheless are aluminium-tolerant. They must also have arisen as the products of natural selection on the aluminium toxic acid soils in which they were grown for agronomic assessment (Briggs and Nyachiro 1988). Similar differences in aluminium tolerance appear to have arisen as the product of natural selection in some wheat cultivars bred in Ohio, USA, that had been grown and selected artificially for agronomic features (but not for aluminium tolerance) on low pH aluminium toxic soils in that State. Similar material from Indiana was poor yielding on these acid soils on which the

Ohio bred varieties performed well (Foy et al. 1974). Again the variation that has been exploited in the evolution of these cultivars must have been present in the mixed genotype land races from which the more genetically uniform cultivars have subsequently been derived.

7.4.2.2 Sorghum

Sorghum (*Sorghum bicolor*) together with pearl millet (*Pennisetum americanum*) constitute the major cereal crops of rain-fed agriculture in the semi-arid tropics, and throughout this zone they are grown on soils with a wide spectrum of soil acidity. Sorghum shows an interesting parallel with the situation found in wheat with respect to aluminium tolerance.

The aluminium tolerance of 1200 sorghum lines of differing origins in Africa was examined by Pitta et al. (1976, 1979). They found that materials from Kenya and Uganda were the most tolerant. In a further examination of 730 lines from the world collection for aluminium tolerance, Gourley (1987) showed that the greatest occurrence of the most highly tolerant category material was from acid soil regions of Uganda (62%) and Kenya (52%). Tanzania (50%) and Zaire (49%) differed little from Kenya, but no information is given about the soils of origin of these materials.

7.4.2.3 Rice

In rice (*Oryza sativa*), an association between soil acidity and the degree of aluminium tolerance has also been reported (Howeler and Cadavid 1976). These authors suggest that the greater aluminium tolerance of upland tall varieties grown on acid soils is probably the result of unconscious/natural selection over a number of generations on those soils. The absence of tolerance in semi-dwarf varieties developed for lowland flooded conditions relates to the absence of selection for tolerance in the non-acid conditions they have been grown in.

The evolution of aluminium tolerance in wheat, sorghum and rice through unconscious/natural selection in response to the appropriate selection pressure (i.e. soil aluminium toxicity) can obviously only have arisen because the necessary genetic variability for tolerance must have been present within the source materials available to farmers, but little or no evidence has yet been obtained about this. Nonetheless, from the number of crop species for which variation in aluminium tolerance has been reported, it would seem that tolerance to aluminium is much more common than tolerance to the heavy metals, as reflected in the admittedly rather limited data available from wild species. This may have important implications for crop improvement in tolerance to aluminium.

7.4.2.4 Maize

Another crop in which the aluminium tolerance of material from Brazil is greater than similar material from the USA is maize (Magnavaca et al. 1987a).

However, there appears to be no direct link between the Al-saturation levels in the soils where the Brazilian lines were developed and their aluminium tolerance. Their tolerance is apparently due to the random fixation in inbred lines of tolerance genes present in the populations from which they were derived. This suggests a high frequency of tolerance genes in those founder populations.

In a preliminary screening of 195 maize (*Zea mays*) populations, Bahia et al. (1978) found considerable variation in tolerance between populations. Four cycles of selection on acid soil, screening 500 half-sib families per year, led to the production of a selected line of the highly tolerant population, Composto Amplo. The variance of this selected line, and its mean aluminium tolerance were, however, lower than those of the original population (Magnavaca et al. 1987a). It was suggested that this could have been due to a negative correlation between tolerance and grain yield, a possibility which requires further investigation in maize, but which is contrary to findings for Brazilian aluminium-tolerant wheats grown on normal soils, where no yield penalty was found (da Silva 1976; Martini et al. 1977). It is worth noting that when they were grown in pure stands, the dry matter yields of zinc-tolerant and normal populations of *Agrostis capillaris* and *Anthoxanthum odoratum* and copper-tolerant and normal *A. capillaris* populations did not differ on normal soil (Cook et al. 1972). However, when corresponding tolerant and non-tolerant populations were grown in mixtures, the normal populations consistently had higher yields.

7.4.2.5 Other Species

Of a number of perennial grasses introduced into Australia, *Phalaris aquatica* is amongst the most productive in the 400–800 mm rainfall areas of the southeast. Soil acidification under subterranean clover cultivation in that area has given rise to increased soil aluminium levels. Following from the report by Helyar and Anderson (1970), who found variation in *P. aquatica* plants growing on aluminium toxic soil, Culvenor et al. (1986a) examined aluminium response in 39 lines of *P. aquatica* and five of *P. arundinacea*. They found significant differences between and within lines of *P. aquatica* between lines of *P. arundinacea*, the latter species being in general the more tolerant.

P. aquatica was also shown to have a relatively high tolerance to soil manganese (Culvenor 1985) and, more importantly, differences in tolerance were found between lines. As a consequence of the widespread occurrence of manganese toxicity in Australia (Scott et al. 1987), the manganese tolerance of pasture legumes has been extensively studied (Helyar 1978). Not unexpectedly, species differed in tolerance, but intra-specific differences in tolerance were recorded where variability was examined at this level. There were wide differences in tolerance among 78 subterranean clover (*Trifolium subterraneum*) lines examined by Evans et al. (1987), who detected tolerant individuals within all the subterranean clover varieties they tested.

Variation in the tolerance of soybean (*Glycine max*) to manganese has been shown (Carter et al. 1975; Heenan and Carter 1975, 1976, 1977), and these differences were subsequently shown to be heritable (Heenan et al. 1981). In lucerne, inter-varietal differences in manganese tolerance were reported by Ouelette and Dessereaux (1958), Dessereaux and Ouelette (1958) and Salisbury and Downes (1982) and also in white clover (*Trifolium repens*) by Vose and Jones (1963). Intra-specific differences in tolerance have also been recorded in wheat, triticale, cotton and flax (Foy 1983a). As is the case with aluminium tolerance in wheat, the cultivars which exhibit the highest manganese tolerance are predominantly of Brazilian origin, with Chilean material also having a high proportion of highly tolerant cultivars. The tolerance is heritable and controlled by two recessive genes (Scott et al. 1987). The same authors reported that selection was able to improve the manganese tolerance of *Brassica campestris* (rape), suggesting polygenic control of the character.

In view of the evidence that boron toxicity in Australia was more widespread than previously thought, Cartwright et al. (1987) examined the response of 25 barley varieties and 150 wheat lines to high boron in natural soil and in solution culture. In both species significant differences in boron tolerance were found, and since some degree of indication of boron tolerance could be obtained from varietal pedigree, the character can be assumed to be heritable, although it does not appear to have a simple genetic basis. As stated earlier, a genetic basis for the phenotypic variation observed for any character is a necessary pre-requisite for any success to be achievable in a breeding programme. The next section will examine this aspect of tolerance to metals.

7.5 The Genetic Basis of Metal Tolerance

7.5.1 Wild Species

The genetic basis of heavy metal tolerance in wild species has been the subject of a recent review (Macnair 1989). This provides a full account of our current knowledge and the author's interpretation of the data available. It emphasises the problems of using root length data because of their high error component and of relative performance data because of the influence which control lengths have on such data. Macnair argues that metal tolerance is, and would be expected to be, controlled by single major genes, with tolerance being dominant, a situation certainly true in *Mimulus guttatus*. Whether this can be argued for other species that have been examined for other metals, and even for the diversity of populations tolerant to a single metal throughout the range of a single species, would seem to be unrealistic. In fact, in one of the earliest examinations of the genetics of tolerance, Wilkins (1960) found both tolerance and susceptibility in *Festuca ovina* to show dominance, depending upon the parents used. Schat and ten Bookum (1992) have reported that presence of copper tolerance in *Silene cucubalis* is controlled by a single major gene for tolerance. However, the degree of tolerance is controlled by a number of hypostatic enhancers. The number of these enhancers which segregate in F_2 material obtained from tolerant × non-tolerant crosses is greater the greater the tolerance of the tolerant parent. How far this simply represents a polygenic control of tolerance is as yet unclear, and the problem of distinguishing enhancer effects from those of polygenes would seem to be insurmountable.

Data from an examination of the genetic basis of zinc tolerance in *A. capillaris* from beneath five electricity pylons (Al-Hiyaly 1989) using a North Carolina Model II crossing programme and root length data rather than relative tolerance (tolerance index) values found both significant additive and non-additive effects. Nonetheless, there was dominance towards tolerance in all four crossing groups examined. The same author, using similar material of the same species, determined realised heritability values ranging from 0.32 to 0.44 from a series of recurrent selection cycles, again suggesting polygenic control of tolerance, a conclusion reached by Nicholls (1977) for zinc tolerance in the same species.

Clearly the number of genes involved requires clarifying, using greater numbers of populations within a species and selection experiments to allow realised heritability estimates to be obtained. Despite these differences in emphasis, it is nonetheless true that tolerance to heavy metals is one of very few instances where the products of natural selection have been subjected to any kind of formal genetic analysis, in spite of the fundamental importance of the genetic control of a character to evolutionary adaptation (Lawrence 1984). Data about the genetic basis of aluminium and manganese tolerance in wild species, or even a clear indication that the character is under genetic control, appear to be lacking in the literature.

7.5.2 Cultivated Species

Whereas in wild species a genetic basis for tolerance to the heavy metals is available and that for aluminium and manganese is lacking, in crop species the situation is quite the reverse. This clearly relates to the importance of producing lines of crop species which are tolerant to aluminium and manganese and the lack of importance of tolerance to heavy metals. The genetics of tolerance of crop species to aluminium and manganese has been considered recently by Devine (1982), Foy (1983a), Woolhouse (1983) and Blum (1988).

The major temperate zone cereals, wheat and barley, show a marked contrast in the genetic architecture of their tolerance to aluminium. In barley, tolerance is controlled by a single dominant gene located on chromosome 4 (Stolen and Anderson 1978). In wheat the situation is considerably more complex. There is evidence that a single dominant gene is responsible for tolerance (Kerridge and Kronstad 1968; Lafever and Campbell 1978). However, from crossing experiments with two tolerant and two sensitive cultivars, the authors concluded that the inheritance pattern was more complex than can be explained by a single gene with incomplete dominance and suggested a complex control of the character.

The validity of this conclusion has been amply shown by more recent work. Using nullisomic-tetrasomic and ditelosomics of the moderately tolerant wheat cultivar Chinese Spring, Aniol and Gustafson (1984) localised a total of seven genes responsible for the moderate tolerance of that cultivar. Five were located on the long arms of chromosomes 6A, 2D, 3D, 4D and 4A (previously assigned to the B genome) one on the short arm of chromosome 7A and a further one on chromosome 7D. Tagaki et al. (1983) also found major gene effects from the long arms of chromosomes 2D and 4D, and a further gene with minor effects on chromosome 2B. More recent evidence (Aniol 1990), again using Chinese Spring, has shown that general tolerance to aluminium is governed by genes on the short arm of chromosome 5A, whilst the genes on D genome chromosomes 2 and 4 are effective only at high aluminium concentrations. Suppression of aluminium tolerance is also suggested by the data from crossing Chinese Spring with a tolerant Brazilian variety, BH 1146, the effect being due to factors on the short arm of chromosome 6B. Aniol (1990) concluded that tolerance is controlled by combinations of (1) genes with major effects, (2) modifiers with minor effects and (3) by further genes which control suppression of tolerance.

The distribution of genes for tolerance may differ between different accessions from different geographical areas, as shown by the findings of Carver et al. (1988) for hard red winter wheats. Should these prove to be genes additional to those detected in Chinese Spring and BH 1146, this could have considerable importance for further breeding for tolerance.

In rye, tolerance to aluminium appears to be under the control of major genes located on chromosomes 3R and 6R, and other genes on chromosome 4R (Aniol and Gustafson 1984). Transferring genes for aluminium tolerance from rye to wheat has appeared an attractive possibility, to allow breeders to increase the tolerance of the recipient species. However, the addition of the rye chromosomes 3R and 4R, and particularly 6R, to the moderately tolerant Chinese Spring failed to effect tolerance to the extent expected from the tolerance of the rye donor (Imperial) although all four substitution lines were more tolerant than the Chinese Spring recipient. Attempts to transfer tolerance to the highly sensitive wheat cultivar Kharkov by addition of chromosomes from Drakold rye failed, no tolerance at all being conferred on the wheat recipient. From these data it is clear that the expression of tolerance genes from rye is dependent upon the impact of genes within the recipient wheat genome background: the degree of enhancement of the tolerance shown is positively correlated to the tolerance of the recipient genotype (Aniol and Gustafson 1984). In none of the substitution lines examined did the increase in aluminium tolerance observed reach that expected from the tolerance of the rye parent.

The amount of genetic information that is available about aluminium tolerance in *Sorghum* is less than for wheat and rye. Pitta et al. (1979) reported that tolerance was controlled by a small number of genes with dominance effects (Bastos 1982). Furlani et al. (1983) examined a number of F_1 sorghum hybrids and again found that tolerance was dominant, but not universally so. Other evidence, however, showed that tolerance was genetically more complex and involved three or more genes (Bastos 1982). From a half diallel analysis involving three tolerant and three susceptible genotypes, Boye-Goni and Macarian (1985) showed that tolerance was predominantly under additive genetic control with some degree of dominance towards tolerance. The narrow sense heritability estimate of 0.78, although based on only six genotypes and a similar estimate of 0.52 from Boye-Goni (1982), suggested that selection should be effective and rapid responses should be obtainable in early generations.

From a series of pair crosses using sorghum, Borgonovi et al. (1987) estimated general and specific combining abilities for aluminium tolerance in solution culture experiments as 0.52 and 0.12, respectively. The former reflected additive variation and additive by additive interactions and the latter non-additive genetic variance. In a recent series of crossing experiments carried out in soil and solution cultures involving the North Carolina Model II and diallel designs, Gourley et al. (1990) found that the type of gene action revealed depended upon the technique used for measurement of tolerance, the genotypes used and the extent of stress to which the material was exposed. Analysis of the North Carolina Model II data for soil-grown plants showed that tolerance of sorghum was predominantly controlled by dominance variation, and narrow sense heritability estimates for shoot and root data respectively, were 0.16, and 0.08 at greater aluminium toxicity levels. By contrast, using the same method of analysis with solution-grown plants, narrow sense heritability values based on shoot and root data, respectively, were 0.72 and 0.65, with partial dominance towards tolerance for shoot growth data, but none for root growth data. Again, from data from a diallel crossing design with progeny plants grown in solution culture, there were significant general and specific combining abilities and the two estimates were of similar magnitude. There appear to be genes in the material examined which have differing degrees of dominance that could be combined with advantage in the expression of tolerance. Nonetheless, the evidence for significant additive genetic variation suggests that considerable advances in aluminium tolerance could be readily achieved.

Aluminium tolerance in maize appears, from data from F_2 and backcross material, to be dominant (Rhue et al. 1978), although tolerance could be divided into four classes suggesting complex genetic control. However, Rhue and Grogan (1977) suggested that tolerance was controlled by a multi-allelic series at a single locus. A further complication arises from the pleiotropic effects of the heat resistant gene Lte2 which confers both heat and aluminium tolerance in maize. From an examination of F_1 and backcross material, Garcia et al. (1979) also concluded that a single major gene, with dominance for tolerance, and the possibility of modifiers, controlled tolerance. Later work, however, using a ten-diallel crossing procedure (Naspolini et al. 1981) showed that general and specific combining ability effects were present, but that general combining ability variances were the higher of the two. Thus, additive genetic effects were also important in their material. Magnavaca et al. (1987b) examined the genetic basis of tolerance in parental, F_1, F_2 and two backcross generations of six maize populations. Additive effects accounted for most of the genetic variation almost twice as much as dominance effects. They also found some evidence that dominance was towards non-tolerance, but it was not consistent. These authors also addressed the possible problems in interpreting data from genetic analyses which are based on different measures of tolerance and different aged plants, and suggest that such inconsistencies may account for the differing results in different studies. There is, nonetheless, clear evidence that both additive and non-additive genetic variation contribute to the genetics of aluminium tolerance in maize.

Howeler and Cadavid (1976) provided data for relative root length values in pair cross progeny of two aluminium-tolerant and two aluminium-susceptible rice cultivars and suggested that tolerance was governed by only a small number of genes (see also Chap. 2.1). Their data suggests that for one of their crosses, dominance for tolerance was complete; for another, that there was partial dominance for tolerance; and for another, additive effects only were involved. From an examination of the F_2 progenies from six tolerant and sensitive rice cultivars, Camargo (1984) recorded partial dominance of sensitivity, and significant maternal effects. However, there were significant additive effects with narrow sense heritability estimates lying between 0.50 and 0.87, depending upon the test aluminium concentration. A considerable range in aluminium tolerance has subsequently been found following mutagenesis of fertilised ova (Chandhry et al. 1986) and amongst second-generation regenerants from seed derived callus (Abrigo et al. 1985), suggesting polygenic control of tolerance.

In *Phalaris* material examined by Culvenor et al. (1986b), differences between highly sensitive and moderately tolerant classes were explained (Culvenor et al. 1986a) on the basis of a two-gene system where tolerance required at least one dominant allele at each locus, with the possibility that modifiers were also involved. Estimates of narrow sense heritability from field-grown parents and their progeny ranged from 0.07 to 0.26, whilst in solution culture the range was from 0.49 to 0.74. Based upon such evidence, substantial gains in tolerance would be expected from selection based upon solution culture screening, but advances using soil screening would seem to be less promising.

It is clear from the studies referred to above, and from evidence about other species, that tolerance to aluminium is under at least some degree of genetic control, in some cases involving single genes with major effect, in others a polygenic system with dominance generally towards tolerance. It also seems clear that some of the differences in the results obtained by workers examining the same species may be due to differences in the method of quantifying tolerance: soil versus solution culture of plants; growing seedlings in normal solution prior to aluminium application or growing seedlings from the outset in conditions of aluminium stress; the degree of tolerance in their experimental material; the age at which tolerance was measured. Given the complexity of the character of tolerance, whether it be heavy metal tolerance in wild species, or aluminium or manganese tolerance in wild or cultivated species, it is difficult to accept that overall control is due to a single major gene. There would seem to be no a priori reason why we should expect the genetic basis of these characters to be the same across different races/lines of the same species, let alone across species, particularly on expectations from Mathers theory of genetic architecture (Mather 1960, 1966, 1973; Lawrence 1984). This would predict the occurrence of a spectrum of genetic architecture from predominantly additive genetic variation to predominantly non-additive variation, depending upon the selection history both of the material and of the character under investigation.

7.6 The Physiological Basis of Tolerance to Metals

Attempts to disentangle the mechanism or mechanisms involved in heavy metal tolerance have in many cases been concerned with tolerance of single cells. How far these mechanisms can be considered to be related to those which operate in such a highly integrated system as the whole plant represents is as yet impossible to predict. Certainly, the evidence that different mechanisms may be responsible for tolerance to salinity in single cells and whole plants suggests that single cell based studies (see also Chap. 5) may not provide a complete understanding of the mechanism of metal tolerances. It also seems logical to suppose that different species may achieve tolerance by the same, partially the same, or quite different mechanisms, and this may also be true of different populations within a single species. In such circumstances it would seem prudent for those who are seeking mechanisms of tolerance to seek them in material which, as far as is possible, differs genetically only in its tolerance.

Two recent reviews have addressed the current status of our knowledge of mechanisms of metal tolerance in the higher plants (Robinson 1989; Verkleij and Schat 1989) and the reader is referred to these articles for further information. It is clear, however, that thus far we are not in a position to be able to provide an explanation of why some genotypes within a species are tolerant, whilst others are not tolerant, and why some species apparently lack the ability to evolve tolerance, yet others clearly can evolve tolerance. I would argue strongly that this does not matter. What is surely much more important is to know more about the effects of the character of metal tolerance, about its genetic basis, and about the extent of available variability. It can only be exploited on the basis of such knowledge.

The physiological basis of tolerance to aluminium has recently been reviewed by Taylor (1988) and the physiology of both aluminium and manganese tolerance were the subject of a review by Foy (1983a). Taylor (1988), with Verkleij and Schat (1989), itemises two main categories of tolerance: exclusion tolerance and internal tolerance, which act in concert. Functions of the cell wall, selective permeability of the plasma membrane, a plant-induced pH barrier at the root soil interface and exudation of chelate ligands are listed by Taylor (1988) as possible external tolerance mechanisms: as possible internal mechanisms he lists chelation in the cytosol, compartmentation in the vacuole, the presence of aluminium-binding proteins and the evolution of aluminium-tolerant enzymes. Whether all of both external and internal mechanisms, or combinations of some of both external and internal mechanisms are involved is unknown. In short, Taylor considered that "a workable hypothesis to explain how plants flourish on Al-rich soils is lacking". The same conclusion would seem to be true of manganese tolerance.

7.7 Synthesis

The requirements for evolutionary change were itemised at the beginning of this chapter. It is clear that for tolerance both to heavy metals, and to aluminium and manganese, these requirements are achieved to a level that is at least acceptable. Firstly, the ability exists to distinguish between plant material which grows on soils having high levels of toxic metals and material of the same species which does not, using simple measures of growth either in soil or in solution cultures. This provides a means whereby selection may be carried out and applied to improve tolerance. The early work on naturally occurring metal-tolerant species/populations (Wilkins 1957; Jowett 1958) laid the foundations for this development. Ultimately, selection could be based upon simple phenotypic differences. Secondly, there is clear evidence of variability in tolerance to aluminium and manganese, some evidence for variation in tolerance to boron and iron, and in a much smaller group of species to heavy metals. Not enough is known about the availability of this variation in different populations and species; but this does not stop the search for it.

Finally, where the genetic basis of the differences in tolerance have been examined, it is clear that to a degree it satisfies the requirements of a breeding programme, i.e. that the variability that has been found is under genetic control. Thus the elements necessary to proceed to the development of cultivars with advanced tolerance through programmes of selection and breeding would seem to be in place.

From the available data it is clear that tolerance to aluminium is much more widespread, and occurs to a higher degree in many more species than does tolerance to the heavy metals. It also seems to be the case that in almost all the cases where screening for aluminium tolerance has been carried out, particularly in the important cereals, wheat, maize, rice and sorghum, the material consists of pure lines – existing cultivars, breeders lines, inbred lines. In these, only free variation (Lerner 1958; Mather 1973) is expressed because of the homozygous nature of such materials in these inbreeding species.

At the same time, contrasting with free variation, is potential variation – that which can be freed only as a consequence of crossing and segregation in subsequent generations. Screening segregating generations – F_2, F_3 etc. – of crosses between pure line having modest tolerances, or even susceptible lines, may produce individuals with significantly advanced tolerance. The intermediate aluminium tolerance of the hexaploid wheats Cimmaron, Payne and Trapper (Carver et al. 1988), all of which were derived from crosses between parents which were rated as being very susceptible to aluminium toxicity, may have its origin from the result of the release of potential variability, manifested as transgressive segregants, in F_2 or subsequent generations. The same argument may explain the different tolerances of the Ohio experimental lines TN1628 and TN1662, sister lines from the same parental lines, the former being aluminium-tolerant, and the latter aluminium-sensitive (Foy et al. 1974). The potential for improving aluminium and manganese tolerance may be much greater than the evidence from examination of existing cultivars of these inbreeding species would suggest. Composite Cross XXXIV, produced by Reid et al. (1980) by crossing a number of tolerant lines, will owe at least some of its variability to the release of potential variability. Where the occurrence of heavy metal tolerance has been sought in wild species, large numbers of individuals have been screened (Gartside and McNeilly 1974; Walley et al. 1974; Symeonidis et al. 1985; Ingram 1987). The large numbers used in these cases ranged from 15000 upwards to 30000, with the objective of isolating highly tolerant individuals with frequencies of 0.001 or less. Screening similar numbers of wild glycophyte species for salinity tolerance has achieved similar success (Ashraf et al. 1986; Al-Khatib 1991).

Such numbers are necessitated by the very low frequencies of the tolerance genes which are being sought. This follows from models predicting the frequency of occurrence of favoured loci in a population. Small numbers of individuals are most likely to contain only the commoner alleles. The predicted frequency (f) of individuals having a single pre-adapted allele at each of n unlinked loci is given by $f = (p^2 + 2pq)^n$. The impact of allele frequency on the expected yield of favoured (adapted) individuals is shown in Table 7.4. In contrast to the work

Table 7.4. Predicted frequency of individuals having at least one favoured allele at each of n unlinked loci in population samples of 10000 individuals

	No. of favoured loci					
	1	2	3	4	5	10
Frequency of favoured allele						
0.001	19	0	0	0	0	0
0.002	40	0	0	0	0	0
0.01	199	4	0	0	0	0
0.05	1200	144	17	2	0	0
0.1	1900	61	68	13	2	0

with heavy metal tolerance, screening for tolerance to aluminium – the most amenable to screening because of the clear phenotypic expression of tolerance in short-term experiments – has used much smaller numbers of individuals; 1795 barley lines (Slootmaker and Reid, in Reid et al. 1980); 686 wheat lines (Tagaki et al. 1983); approximately 1360 rice lines (Martinez and Sarkarung 1984). These represent but a small proportion of the available germplasm in these inbreeding species, which may contain much potential variability.

It is interesting to speculate on the extent to which tolerance genes with very low frequency would be uncovered by screening respectably large numbers of individuals such as segregating generations of inbreeding species. Within naturally outcrossing species, variability at the intra-cultivar/breeders line level in more tolerant material could provide useful sources of high tolerance. The data about variation in aluminium tolerance in white clover (Vose and Jones 1963), in soybean (Carter et al. 1975) and in lucerne by Dessereaux and Ouelette (1958), and particularly in that species by Devine et al. (1976), who carried out recurrent selection with considerable success, illustrate the potential for improvement in tolerance that exists.

Despite the obvious success that has been achieved by plant breeders selecting for improvements in a very broad spectrum of characters for which no "mechanism" is known, there seems to be a fairly broad consensus amongst plant physiologists that knowledge of "a mechanism" will provide breeders with a proper means for selecting for tolerance to salinity, metal toxicity, drought and temperature sensitivity. Much research seeking to understanding the physiological basis (i.e. "the mechanism") of tolerance to aluminium and manganese has been aimed at providing the breeder with a means to screen for tolerance with precision. Clearly our understanding of possible mechanisms is not at the stage at which the use of such information is a possibility.

A major problem in reaching this goal is the very variability that is the overwhelming characteristic of living material. Different species may well achieve the same degree of tolerance to the same stress by different mechanisms and it is not inconceivable that different members of the same species may also achieve tolerance by different mechanisms. Different tolerance mechanisms may function at different stages in the life of the plant and make selection for overall tolerance extremely difficult. Probably the greatest constraint lies in the fact that in order to be of use to the breeder, any method of determining the mechanism must be non-destructive for any material which is not available as isogenic lines, or which can be successfully cloned prior to examination.

In promoting mechanisms as a basis for selection for tolerance, it has to be remembered that a proper screening will involve not tens or hundreds of individuals, but tens of hundreds or thousands of individuals, so that any method would need to cope routinely and cheaply with such large numbers. From a knowledge of mechanisms it should ultimately be possible to identify alleles responsible for them. In turn, this should lead to an indication of the effects of those different alleles on phenotype. Based on such information it should then be possible to breed lines adapted to specific toxicities. Whether these are reasonable goals, justifiable in terms of breeding strategy rather than on purely academic grounds, remains to be seen, particularly in view of the potential that seems to exist for improvements in tolerance using simple phenotypic selection based on the characters outlined above.

Currently it would seem that the plant breeder is likely to be of more immediate use to the physiologist than vice versa. Selected lines, isogenic at all loci other than those for tolerance, or of similar genetic background, would provide the correct material for seeking mechanisms, and could be readily provided by the breeder, since possible confusing effects of genes at loci other than those involved in tolerance would then be eliminated.

References

Abrigo WM, Novero AU, Coronel VP, Cabuslay GS, Blanco LC, Parao FT, Yoshido AS (1985) Somatic cell culture at IRRI. Biotechnology in International Agricultural Research. IRRI, Manila, Philippines, p 149

Al- Hiyaly SAK (1989) Evolution of zinc tolerance under electricity pylons. PhD Thesis, University of Liverpool, Liverpool

Al-Khatib M (1991) Salinity tolerance breeding in lucerne. PhD Thesis, University of Liverpool, Liverpool

Allen WR, Sheppard PM (1971) Copper tolerance in some Californian populations of the monkey flower, *Mimulus guttatus*. Proc R Soc Lond Ser B 177: 177–196

Andrew CS, Hegarty MP (1969) Comparative responses to manganese excess of eight tropical and four temperate legumes. Aust J Agric Res 20: 687–696

Aniol A (1990) Genetics of tolerance to aluminium in wheat (*Triticum aestivum* L. Thell). Plant Soil 123: 223–227

Aniol A, Gustafson JP (1984) Chromosome location of genes controlling aluminium tolerance in wheat, rye, and triticale. Can J Genet Cytol 26: 701–705

Ashraf M, McNeilly T, Bradshaw AD (1986) The potential for evolution of salt (NaCl) tolerance in seven grass species. New Phytol 103: 299–309

Bahia AFC, Franca GE, Pitta GVE, Magnavaca R, Mendes JF, Bahia FGFTC, Pereira P (1978) Evaluation of corn inbred lines and populations in soil acidity conditions. XI Annu Brasilian Maize Sorghum Conf, Piracicaba SP, Brasil, pp 51–58 (in Portugese)

Baker AJM (1978) Ecophysiological aspects of zinc tolerance in *Silene maritima* With. New Phytol 80: 635–642

Baker AJM, Brooks RR (1989) Terrestrial higher plants which hyper-accumulate metallic elements – a review of their distribution, ecology, and phytochemistry. Biorecovery 1: 81–126

Baker AJM, Walker PL (1989) Physiological responses of plants to heavy metals and the quantification of tolerance and toxicity. Chem Spec Bioavail 1: 7–17

Baligar VC, Kinraide TB, Wright RJ, Bennett OL (1987) Al effects on growth and P, Ca and Mg uptake efficiency in red clover cultivars. J Plant Nutr 10: 131–1137

Bastos CR (1982) Inheritance study of aluminium tolerance in sorghum in nutrient culture. PhD Thesis, Mississippi State University, Mississippi State

Baumeister W (1954) Über den Einfluss des Zinks bei *Silene inflata* Smith. I Mitteilung. Ber Dtsch Bot Ges 67: 205–213

Baumeister W, Burghardt H (1956) Über den Einfluss des Zinks bei *Silene inflata* (With) II Mitteilung: CO_2-Assimilation und Pigmentgehalt. Ber Dtsch Bot Ges 69: 161–168

Blum A (1988) Plant breeding for stress environments. CRC, Boca Raton, 223 pp

Borgonovi RA, Schaffert RE, Pitta GVE, Magnavaca R, Alves VMC (1987) Aluminium tolerance in *Sorghum*. In: Gabelman WH, Loughman BC (eds) Genetic aspects of plant mineral nutrition. Nijhoff, Dordrecht, pp 213–221
Boye-Goni SR (1982) Combining ability and inheritance of aluminium tolerance in grain sorghum [*Sorghum bicolor* (L.) Moench]. PhD Thesis, University of Arizona, Tucson
Boye-Goni SR, Macarian V (1985) Diallel analysis of aluminium tolerance in selected lines of grain sorghum. Crop Sci 25: 749–752
Bradshaw AD, McNeilly T (1981) Evolution and pollution. Arnold, London
Briggs KE, Nyachiro JM (1988) Genetic variation for aluminium tolerance in Kenyan wheat cultivars. Commum Soil Sci Plant Anal 19: 1273–1284
Broker W (1963) Genetisch-physiologische Untersuchungen über die Zinkverträglichkeit von *Silene inflata* Sm. Flora 153: 122–156
Brooks RR, Malaisse F (1989) Mineral enriched sites in South Central Africa. In: Shaw AJ (ed) Heavy metal tolerance in plants: evolutionary aspects. CRC, Boca Raton, p 53
Camargo CEO (1984) Genetic evidence of aluminium tolerance in rice. Bragantia 43: 95–110 (in Portuguese)
Campbell AT, Nuernberg NJ, Foy CD (1989) Differential response of alfalfa to aluminium stress. J Plant Nutr 12: 291–305
Campbell LG, Lafever TN (1976) Correlation of field and nutrient culture techniques of screening wheat for aluminium tolerance. In: Wright M (ed) Plant adaptation to mineral stress in problem soils. Cornell Univ Agric Stn, Ithaca, NY, pp 277–286
Cardus JR (1987) Intraspecific variation for tolerance to aluminium toxicity in white clover. J Plant Nutr 10: 821–830
Carter OG, Rose IA, Reading PF (1975) Variation in susceptibility to manganese in 30 soybean lines. Crop Sci 15: 730–732
Cartwright B, Rathgen AJ, Sparrow DHB, Paull JG, Zarcinas BA (1987) Boron tolerance in Australian varieties of wheat and barley. In: Gabelman HW, Loughman BC (eds) Genetic aspects of plant mineral nutrition. Nijhoff, Dordrecht, pp 131–151
Carver BF, Inskeep WP, Wilson NP, Westerman RL (1988) Seedling tolerance to aluminium toxicity in hard red winter wheat germplasm. Crop Sci 28: 463–467
Chandhry MA, Yoshida S, Vergara BS (1986) Induced mutations for aluminium tolerance after N-methyl-N-nitrosourea treatment of fertilized egg cells in rice. Environ Exp Bot 27: 37–43
Clark RB (1982) Plant response to mineral element toxicity and deficiency. In: Christiansen MN, Lewis CF (eds) Breeding plants for less favourable environments. Wiley, New York, pp 71–142
Clarkson DT (1966) Aluminium tolerance in species within the genus *Agrostis*. J Ecol 54: 167–178
Clymo RS (1962) An experimental approach to part of the calcicole problem. J Ecol 50: 707–731
Cook SCA, Lefebre C, McNeilly T (1972) Competition between metal tolerant and normal plant populations on normal soil. Evolution 26: 366–372
Culvenor RA (1985) Tolerance of *Phalaris aquatica* L. populations and some agricultural species, and the effect of aluminium on manganese tolerance of *P. aquatica*. Aust J Agric Res 36: 695–708
Culvenor RA, Oram RN, Fazekas de Groth C (1986a) Variation in tolerance in *Phalaris aquatica* L. and a related species to aluminium in nutrient solution and soil. Aust J Agric Res 37: 383–395
Culvenor RA, Oram RN, Wood JT (1986b) Inheritance of aluminium tolerance in *Phalaris aquatica* L. Aust J Agric Res 37: 397–408
Dessereaux L, Ouelette CJ (1958) Tolerance of alfalfa to manganese toxicity in sand culture. Can J Soil Sci 38: 8–13
Devine TE (1982) Genetic fitting of crops to problem soils. In: Christiansen MN, Lewis CF (eds) Breeding plants for less favourable environments. Wiley, New York, pp 143–173

Devine TE, Foy CD, Fleming AL, Hanson TA, Campbell TA, McMurtrey JE, Schwartz JW (1976) Development of alfalfa strains with differential tolerance to aluminium toxicity. Plant Soil 44: 73–79

Duncan RR (1988) Sequential development of acid soil-tolerant sorghum genotypes under field stress conditions. Commun Soil Sci Plant Anal 19: 1295–1305

Evans J, Scott BL, Lill WJ (1987) Manganese tolerance in subterranean clover (*Trifolium subterraneum* L.) genotypes grown with nitrate or symbiotic nitrogen. Plant Soil, 97: 207–215

Foy CD (1974) Effects of aluminium in plant growth. In: Clarkson EW (ed) The plant root and its environment. University Press of Virginia, Charlottesville, pp 57–97

Foy CD (1983a) Plant adaptation to mineral stress in problem soils. Iowa State J Res 57: 339–354

Foy CD (1983b) The physiology of plant adaptation to mineral stress. Iowa State J Res 57: 355–391

Foy CD (1988) Plant adaptation to acid, aluminium-toxic soils. Commun Soil Sci Plant Anal 19: 959–987

Foy CD, Lafaver NH, Schwartz JW, Fleming AL (1974) Aluminium tolerance of wheat cultivars related to region of origin. Agron J 66: 751–758

Furlani PR, Clark RB (1987) Plant traits for evaluation of responses of sorghum genotypes to aluminium. In: Gabelman HW, Loughman BC (eds) Genetic aspects of plant nutrition. Nijhoff, The Hague, pp 247–254

Furlani PR, Clark RB, Ross WM, Maranville JW (1983) Variability and genetic control of aluminium tolerance in sorghum genotypes. In: Saric MR, Loughman BC (eds) Genetic aspects of plant nutrition. Nijhoff, The Hague, pp 453–461

Garcia O, da Silva WJ, Massei MAS (1979) An efficient method for screening maize inbreds for aluminium tolerance. Maydica 24: 75–82

Gartside DW, McNeilly T (1974) The potential for evolution of heavy metal tolerance in plants III. Copper tolerance in normal populations of different species. Heredity 32: 335–348

Gourley LM (1987) Identifying aluminium tolerance in sorghum genotypes grown on tropical acid soils. In: Gabelman HW, Loughman BC (eds) Genetic aspects of plant mineral nutrition. Nijhoff, Dordrecht, pp 89–98

Gourley LM, Rogers SA, Ruiz-Gomez C, Clark RB (1990) Genetic aspects of aluminium tolerance in sorghum. Plant Soil 123: 211–216

Gries B (1966) Zellphysiologische Untersuchungen über die Zinkresistenz bei Galmeiokotypen und Normalformen von *Silene cucubalis* Wib. Flora 156: 271–290

Grime JP, Hodgson JG (1969) An investigation of the ecological significance of lime chlorosis by means of large-scale comparative experiments. In: Rorison IH (ed) Ecological aspects of the mineral nutrition of plants. Blackwell, Oxford, pp 67–99

Heenan DP, Carter OG (1975) Response of two soybean cultivars to manganese toxicity as affected by pH and calcium levels. Aust J Agric Res 26: 967–974

Heenan DP, Carter OG (1976) Tolerance of soybean cultivars to manganese toxicity. Crop Sci 16: 389–391

Heenan DP, Carter OG (1977) Influence of temperature on the expression of manganese tolerance by two soybean varieties. Plant Soil 47: 219–227

Heenan DP, Campbell LC, Carter OG (1981) Inheritance of tolerance to high manganese supply in soybeans. Crop Sci 21: 626–627

Helyar KR (1978) Effects of aluminium and manganese toxicity on legume growth. In: Andrews CS, Kamprath EJ (eds) Mineral nutrition of legumes in tropical and subtropical soils. CSIRO, Melbourne, pp 207–231

Helyar KR, Anderson AJ (1970) Some effects of the soil pH on different species and on the soil solution for a soil high in exchangeable aluminium. Proc XI Int Grassland Congr, Surfer's Paradise, Queensland University. Queensland Press, Brisbane, pp 431–434

Hill PR, Alrichs JL, Ejeta G (1989) Rapid evaluation of sorghum for aluminium tolerance. Plant Soil 114: 85–90

Hoffer GN, Carr RH (1923) Accumulation of aluminium and iron compounds in corn plants and its probable relation to root rots. J Agric Res 23: 801–824
Horst WJ (1987) Aluminium tolerance and calcium efficiency in cowpea genotypes. J Plant Nutr 10: 1121–1129
Howeler RH, Cadavid LF (1976) Screening for rice cultivars for tolerance to aluminium toxicity in nutrient solutions compared with a field screening method. Agron J 68: 551–555
Humphreys MO, Nicholls MK (1984) Relationships between tolerance to heavy metals in *Agrostis capillaris* L. (*A. tenuis* Sibth.) New Phytol 98: 177–190
Ingram C (1987) The evolutionary basis of ecological amplitude of plant species. PhD Thesis, University of Liverpool, Liverpool
Ingrouille MJ, Smirnoff N (1986) *Thalaspi caerulescens* J & C Presl (*T. alpestre* L.) in Britain. New Phytol 102: 219–233
Ittu G, Saulescu NN (1988) Ameliorarea tolerantei la toxicitatea de alumini la triticale. Probl Gen Teor si Aplic 20: 67–74
Jowett D (1958) Populations of *Agrostis* spp tolerant to heavy metals. Nature 182: 816–817
Kerridge PC, Kronstad WE (1968) Evidence of genetic resistance to aluminium toxicity in wheat (*Triticum aestivum* Vill. Host). Agron J 60: 710–711
Kruckeberg AR (1984) California serpentines. University of California Press, Berkeley
Lafever HN, Campbell LG (1978) Inheritance of aluminium tolerance in wheat. Can J Genet Cytol 20: 355–364
Lawrence MJ (1984) The genetic analysis of ecological traits. In: Shorrocks B (ed) Evolutionary ecology. Blackwell, London pp 27–64
Lefèbvre C (1967) Etude de la position des populations d'*Armeria calaminaires* de Belgique et des environs d'Aix la Chapelle par rapport à des types alpines et maritimes d'*Armeria maritima* (Mill) Willd. Bull Soc R Bot Belg 100: 231–239
Lefèbvre C (1968) Note sur un indice de tolerance chez des populations d'*Armeria maritima* (Mill) Willd. Bull Soc R Bot Belg 102: 5
Lerner IM (1958) The genetic basis of selection. Wiley, New York
Little R (1988) Plant soil interaction at low pH. Problem solving – the genetic approach. Commun Soil Sci Plant Anal 19: 1239–1257
MacLean AA, Chiasson TC (1966) Differential performance of two barley cultivars to varying aluminium concentrations. Can J Soil Sci 46: 147–153
Macnair MR (1989) The genetics of metal tolerance in natural populations. In: Shaw AJ (ed) Heavy metal tolerance in plants: evolutionary aspects. CRC, Boca Raton, pp 235–253
Magnavaca R, Gardner GO, Clark RB (1987a) Evaluation of inbred maize lines for aluminium tolerance in nutrient solution. In: Gabelman HW, Loughman BC (eds) Genetic aspects of plant mineral nutrition. Nijhoff, Dordrecht, pp 255–265
Magnavaca R, Gardner GO, Clark RB (1987b) Inheritance of aluminium tolerance in maize. In: Gabelman HW, Loughman BC (eds) Genetic aspects of plant mineral nutrition. Nijhoff, Dordrecht, pp 201–212
Martinez CP, Sarkarung S (1984) Tolerance to aluminium toxicity in upland rice for acid soils. In: Sorghum for acid soils. Proc Worksh Evaluating sorghum for tolerance to aluminium toxic tropical soils in Latin America. INTSORMIL-TCRISAT-CIAT, Cali, Colombia, pp 187–196
Martini JA, Kochann RA, Gomes EP, Langer F (1977) Response of wheat cultivars to liming in some acid high aluminium oxisols of Rio Grande del Sol, Brazil. Agron J 69: 612–616
Mather K (1960) Evolution in polygenic systems. Evol e Genetica, Acad Nazi Lincei, Rome, pp 131–152
Mather K (1966) Variability and selection. Proc R Soc Lond Ser B 164: 328–340
Mather K (1973) Genetical structure of populations. Chapman and Hall, London
McLean AA, Gilbert BE (1927) The relative aluminium tolerance of crop plants. Soil Sci 24: 163–174

McNeilly T (1968) Evolution in closely adjacent populations III *Agrostis tenuis* on a small copper mine. Heredity 23: 99–108
McNeilly T (1982) A rapid method for screening barley for aluminium tolerance. Euphytica 31: 237–239
Moore DP, Kronstad WE, Metzger RJ (1976) Screening for aluminium tolerance. In: Wright MJ (ed) Plant adaptation to mineral stress in problem soils. Cornell University Press, Ithaca, NY, p 287
Morrey DR, Blackwill K, Blackwill MJ (1989) Studies on serpentine flora: Preliminary analyses of soils and vegetation associated with serpentine rock formations in the South-Eastern Transvaal. S Afr J Bot 55: 171–177
Naspolini V, Bahia AFC, Viana RT, Gama EFG (1981) Performance of inbreds and single crosses in corn in soils under cerrado vegetation. Cienc Cult 33: 722–727
Neeling AJ de, Ernst WHO (1986) Response of an acidic and a calcareous population of *Chamaenerion angustifolium* (L.) to iron, manganese and aluminium. Flora 178: 85–92
Nicholls MK (1977) Ecological genetics of copper tolerance in *Agrostis tenuis* Sibth. PhD Thesis, University of Liverpool, Liverpool
Ouelette CJ, Dessereaux L (1958) Chemical composition of alfalfa as related to degree of tolerance to manganese and aluminium. Can J Plant Sci 38: 206–214
Pegtel DM (1986) Responses of plants to aluminium, manganese, and iron, with particular reference to *Succisa pratensis* Moench. Plant Soil 93: 43–55
Pitta GVE, Trevisian WL, Schaffert RE, de Franca GE, Bahia AFC (1976) Evaluation of *Sorghum* lines under high acidity conditions. In: Geres GC (ed) Proc XIth Brazilian Maize Sorghum Rev, Piracicaba, Brazil pp 553–557
Pitta GVE, Schaffert RE, Borgonovi RA, Vasconsellos CA, Bahia AFC, Oliviera AC (1979) Evaluation of sorghum lines to high soil acidity conditions. In: dos Santos AF (ed) Proc XIIth Brazilian Corn Sorghum Res Conf, Gioana, Brazil, p 217
Polle E, Konsak CF, Kittrick JA (1978) Visual detection of aluminium tolerance levels in wheat by haematoxylin staining of seedling roots. Crop Sci 18: 823–827
Polson DE, Adams MW (1970) Differential response of navy beans (*Phaseolus vulgaris*) to zinc. Differential growth and elemental composition at excessive zinc levels. Agron J 62: 557–560
Prat S (1934) Die Erblichkeit der Resistenz gegen Kupfer. Ber Dtsch Bot Ges 102: 65–67
Ramarkrishnan PS (1968) Nutritional requirements of the edaphic ecotypes of *Melilotus alba* Medic. II Aluminium and manganese. New Phytol 67: 301–308
Ramarkrishnan PS (1969) Nutritional factors influencing the distribution of the calcareous and acidic populations in *Hypericum perforatum*. Can J Bot 47: 175–181
Rechcigl JE, Reneau RB, Zelazny LW (1988) Soil solution aluminium as a measure of aluminium toxicity to alfalfa in acid soils. Commun Soil Sci Plant Anal 19: 989–1001
Reid DA (1976) Genetic potential for solving problems of soil mineral stress: Aluminium and manganese tolerances in the cereal grains. In: Wright MJ (ed) Plant adaptation to mineral stress in problem soils. Cornell University Press, Ithaca, NY, pp 55–64
Reid DA, Slootmaker La, Craddock JC (1980) Registration of Composite Cross XXXIV. Crop Sci 20: 416–417
Repp G (1963) Die Kupferresistenz des Protoplasmas höherer Pflanzen auf Kupfererzboden. Protoplasma 57: 643–659
Rhue RD, Grogan CO (1977) Screening corn for aluminium tolerance using different calcium and magnesium concentrations. Agron J 69: 775–760
Rhue RD, Grogan CO, Stockmeyer EW, Everett HL (1978) Genetic control of aluminium tolerance in corn. Crop Sci 18: 1063–1067
Richards RA (1983) Should selection for yield in saline regions be made on saline or non-saline soils? Euphytica 32: 431–438
Richards RA, Dennett CW (1980) Variation in salt concentration in a wheat field. University of California Co-operative Extension. Soil Water 44: 8–9
Robinson NJ (1989) Metal binding polypeptides in plants. In: Shaw AJ (ed) Heavy metal tolerance in plants: evolutionary aspects. CRC, Boca Raton, pp 195–214
Rorison IH (1960) Some experimental aspects of the calcicole-calcifuge problem II. The effects of mineral nutrition on seedling growth in nutrient solution. J Ecol 48: 679–688

Rorison IH (1969) Ecological inferences from laboratory experiments on mineral nutrition. In: Rorison IH (ed) Ecological aspects of the mineral nutrition of plants. Blackwell, Oxford, pp 155–175
Salisbury PA, Downes RW (1982) Breeding lucerne for tolerance to acid soils. In: Yates JJ (ed) Proc 2nd Aust Agron Conf, Wagga, NSW, Australian Society for Agronomy, Parkville, Victoria, pp 339–346
Schat H, ten Bookum WM (1992) Genetic control of copper tolerance in *Silene vulgaris*. Heredity 68: 219–229
Scott BJ, Fisher JA (1989) Selection of genotypes tolerant of aluminium and manganese. In: Robson AD (ed) Soil acidity and plant growth. Academic Press, Mattickville, Australia, pp 167–203
Scott BJ, Burke DR, Bostrom TE (1987) Australian research on tolerance to toxic manganese. In: Gabelmann HW, Loughman BC (eds) Genetic aspects of plant mineral nutrition. Nijhof, Dordrecht, pp 153–163
Silva AR da (1976) Application of the plant genetic approach to wheat culture in Brazil. In: Wright EJ (ed) Plant adaptation to mineral stress in problem soils. Cornell University Press, Ithaca, NY, pp 223–231
Stolen O, Anderson S (1978) Inheritance of tolerance to low soil pH in barley. Hereditas 88: 101–105
Symeonidis L, McNeilly T, Bradshaw AD (1985) Interpopulation variation in tolerance to cadmium, copper, lead, nickel, and zinc in nine populations of *Agrostis capillaris* (L.). New Phytol 101: 317–324
Tagaki H, Namai H, Murakami K (1983) Exploration of aluminium tolerant genes in wheat. Proc 6th Int Wheat Genetics Symp. Maruzen, Kyoto, Japan, p 143
Taylor GJ (1988) The physiology of aluminium tolerance in higher plants. Commun Soil Sci Plant Anal 19: 1179–1194
Thompson J (1987) Population biology of *Anthoxanthum odoratum, Plantago lanceolata*, and *Rumex acetosa* on zinc and lead mine spoil. PhD Thesis, University of Liverpool, Liverpool
Verkleij JAK Schat H (1989) Mechanisms of metal tolerance in higher plants. In: Shaw AJ (ed) Heavy metal tolerance in plants: evolutionary aspects. CRC, Boca Raton, pp 179–193
Vose PB, Jones DG (1963) The interaction of manganese and calcium on nodulation and growth in three varieties of *Trifolium repens*. Plant Soil 18: 372–385
Wachsmann C (1961) Wasserkultur zur Wirkung von Blei, Kupfer und Zink auf die Gartenform und Schwermetallbiotypen von *Silene inflata*. Thesis, University of Münster, Münster
Walley Ka, Khan MS, Bradshaw AD (1974) The potential for evolution of heavy metal tolerance in plants I. Copper and zinc tolerance in *Agrostis tenuis*. Heredity 32: 309–319
Walsh LM, Steevens DR, Siebel HD, Weis GE (1972) Effect of high rates of zinc on several crops grown on an irrigated plainfield sand. Commun Soil Sci Plant Anal 3: 187–195
White MC, Decker AM, Chaney RL (1979) Differential cultivar tolerance in soybean to soil zinc I. Range of cultivar response. Crop Sci 71: 121–125
Wild H (1964) The endemic species of the Chimanimani Mountains and their significance. Kirkia 4: 125–157
Wild H (1965) The flora of the Great Dyke of Southern Rhodesia with special reference to the serpentine soils. Kirkia 5: 49–86
Wild H, Bradshaw AD (1977) The evolutionary effects of metalliferous and other anomalous soils in south central Africa. Evolution 31: 282–293
Wilkins DA (1957) A technique for the measurement of lead tolerance in plants. Nature 180: 37–38
Wilkins DA (1960) The measurement and genetic analysis of lead tolerance in *Festuca ovina*. Ann Rep Scott Plant Breed Stn 1960: 85–98
Williams CH (1980) Soil acidification under clover pasture. Aust J Exp Agric Anim Husb 20: 561–567

Woolhouse H (1983) Toxicity and tolerance in the response of plants to metals. In: Lange OL, Nobel PS, Osman CB, Ziegler H (eds) Physiological plant ecology. III. Response to the chemical and biological environment. Springer Berlin, Heidelberg NewYork, p 254

Wright MJ (1976) Plant adaptation to mineral stress in problem soils. Cornell University Press, Ithaca, NY, 420 pp

Chapter 8
Micronutrient Toxicities and Deficiencies in Rice

H. U. Neue and R. S. Lantin

8.1 Introduction

The area of rice harvested in the world increased during the past 40 years by 41% but rough rice production has increased by 304% (IRRI 1991). It is estimated that the world's annual rough rice production must increase still further, from 519 million tons in 1990 to 758 million tons by 2020 if food production per caput is to be maintained (IRRI 1989). In Asia, where arable land is scarce and population pressure is high, most of the targeted production must come from existing rice land. Much of this land suffers from soil nutrient problems, which will need to be overcome if productivity is to be increased.

Micronutrient deficiencies and toxicities have been researched for more than a century. Gris (1844, 1847) identified Fe-deficiency-induced chlorosis and corrected the deficiency through foliar application and soil application of Fe salts. Conclusive evidence that copper and molybdenum are essential micronutrients for plants was established in the 1930s (Sommer 1931; Arnon and Stout 1939). Chlorine was reported as beneficial in 1862 (Nobbe 1865; quoted in Bergmann 1988) but not until the early 1950s was it proven to be an essential nutrient (Broyer et al. 1954).

Iron deficiency was one of the first micronutrient disorders recognised in rice. As early as 1930, rice was known to be more susceptible to Fe deficiency than other cereals: the highly reductive soil conditions in rice soils were thought to be the primary cause for most "physiological diseases" in early reports (Baba and Harada 1954). Although micronutrients have been intensively researched in the last decades, our understanding of their functions in the soil-plant system is still incomplete. Intensification of crop production has increased the occurrence of nutritional disorders because of lack of knowledge and inappropriate management technologies.

The most common micronutrient disorders of rice are Zn deficiency, Fe deficiency and B toxicity in wetland rice, and Fe deficiency, B deficiency and Mn toxicity in upland rice. Copper deficiency may occur together with Zn deficiency, though both Mn and Mo deficiencies are rare. Chlorine deficiency has not been reported but Cl toxicity occurs in saline soils.

Nutrient deficiencies may be corrected by applying the element to the soil or to the plant. Nutrient toxicities and imbalances are often more difficult to

Monographs on Theoretical and Applied Genetics, Vol. 21
Ed. by A. R. Yeo and T. J. Flowers

remedy but may be alleviated by crop and soil management. Most rice farmers do not, however, have even the resources (standard fertilisers) to correct deficiencies of N, P and K. The use of tolerant cultivars in areas of moderate nutrient disorders is a remedy upon which the farmer of marginal land relies heavily. In this chapter we focus on the occurrence and effects of micronutrient disorders (concerned with Fe, Mn, Zn, Cu, B and Mo) in plants, with special focus on rice.

8.2 Iron

Since iron is highly immobile within the plant and is not appreciably translocated from older to younger tissue, plants have continuously to take up Fe. Iron is the most abundant micronutrient, it has the ability to form stable compounds with sulphur and oxygen and has the capacity to occur in two valence states, which makes it an excellent oxidation/reduction indicator. Iron is present in all soils in primary minerals, clays and hydroxides and it predominates in highly weathered soils. In aerobic soils, the ionic forms (as Fe^{3+} and Fe^{2+}) contribute very little to the total Fe at physiological pH (Lindsay 1974). Inorganic solid phases, mainly ferric oxides, establish solubility limits for uptake and removal of iron by plants. In soil solutions of aerobic soils Fe concentrations should theoretically be less than 0.001 mg/l but concentrations in the range of 0.01–0.1 mg/l are found in practice because of colloidal Fe and organic complexes or chelates (Krauskopf 1972). Chelators may be organic substrates originating as root exudates, soil organic matter and microorganisms (Mengel and Kirkby 1978). Organic matter also influences Fe solubilisation indirectly through the enhancement of soil reduction (Lindsay 1991).

On flooding of the soil, iron is reduced and its concentration may increase to values of 0.1 to 6000 mg/l. The increase in concentration of water-soluble Fe following flooding has been described by the following equations (Ponnamperuma 1972):

$$Fe(OH)_3 + 3H^+ + e = Fe^{2+} + 3\,H_2O \qquad (8.1)$$

$$\text{with} \quad Eh = 1.06 - 0.059 \log Fe^{2+} - 0.177\ pH \qquad (8.2)$$

$$\text{or} \quad pH = 17.87 + pFe^{2+} - 3\ pH, \qquad (8.3)$$

where Eh is redox potential in volts, Fe^{2+} is the activity of water-soluble Fe^{2+}, and pE is the negative logarithm of the electron activity equal to Eh/0.0591.

Siderite ($FeCO_3$). ferrosic hydroxide [$Fe_3(OH)_8$], goethite (FeOOH), ferrihydrite, amorphous ferric hydroxide, hydrotroilite ($FeS \cdot nH_2O$), machinawite (FeS), pyrite (FeS_2) and vivianite [$Fe_3(PO_4)_2$] are the minerals that control the solubility of Fe in flooded soils (Ponnamperuma 1972; Schwab and Lindsay 1983; Neue and Bloom 1989). Lindsay (1991) considers that elevated levels of soluble Fe for extended periods correspond to ferrosic hydroxides. Iron is transported to the root surface as a ferrated chelate or as the ferrous ion by diffusion and mass flow. Although various soil, microbial and synthetic chelates might be similarly effective in mobilising ferric Fe, ferrated phytosiderophores (FeIII PS) are the most important by two to three orders of magnitude so far as uptake by the roots of graminaceous species is concerned (Marschner et al. 1987, 1990) because of the presence of a specific transport system. Phytosiderophores are non-protein amino acids that complex with sparingly soluble Fe(III) and are taken up as Fe-phytosiderophores at the plasma membrane of root cells (Takagi 1976; Marschner et al. 1986). Dicotyledonous species and non-graminaceous monocotyledons also mobilise ferric iron by chelation with root exudates such as caffeic acid, but chelates other than phytosiderophores break up at the plasmalemma and the Fe(III) is reduced to Fe^{2+} before being taken up (Chaney et al. 1972; Mengel and Kirkby 1978).

At low concentrations Fe is taken up, independent of water consumption, by active absorption (Tadano and Yoshida 1978). According to Bergmann (1988) and Römheld and Marschner (1986), plants counteract low availability of Fe by both non-specific and specific mechanisms. Non-specific mechanisms include lowering the pH in the rhizosphere (through cation uptake), exudation of organic acids, and exudation of other metabolic substrates that indirectly influence the pH, redox and chelating power of the rhizosphere via the mediation of microorganisms. Specific mechanisms in rice and other graminaceous species include increased proton exudation, enhanced Fe(III) reduction at the plasmalemma, increased exudation of phenolic compounds (all of which are of minor importance), and increased exudation of phytosiderophores. Phytosiderophores may differ not only between species but also between cultivars (Zhang et al. 1989). Mugineic acid and aveneic acid have been identified as phytosiderophores in rice (Sugiura et al. 1981; Marschner 1986). Compared to wheat, rice increases exudation of phytosiderophores only slightly when Fe becomes deficient (Takagi 1976), which may be related to its natural habitat of flooded soils in which the availability of Fe is generally high.

Deficiencies of Zn or Cu also result in increased exudation of phytosiderophores irrespective of the nutritional status of Fe (Treeby et al. 1989). Enhanced release of phytosiderophores can increase the uptake of one micronutrient while decreasing others (Sugiura et al. 1981; Amberger et al. 1982). Other metals, in the order $Cu^{2+} > Ni^{2+} > Co^{2+} > Zn^{2+} > Cr^{2+} > Mn^{2+} > Ca^{2+}$, can exchange with Fe in chelates and, as a result, may severely hamper Fe uptake and transport in the plant. Furthermore, high concentrations of phosphate may result in the precipitation of Fe in the rhizosphere, root and shoot.

Iron is translocated in the xylem in the form of ferric citrate (Brown 1978). In the shoot and leaves the ferric citrate can be reduced to Fe^{2+} ions via ferrous citrate by light of wavelength < 500 nm. The stability constant for ferric citrate (at 2.5×10^{11}) is much higher than that of ferrous citrate (2.4×10^4).

The metabolic importance of iron results from its tendency to form coordination complexes or chelates and its ability to undergo oxidation–reduction reactions. Iron is present in the prosthetic group of haem proteins (cytochromes, catalases and peroxidases) and non-haem proteins (ferredoxin). Iron has roles in photosynthesis and respiration, chlorophyll formation, oxidation–reduction reactions, protein metabolism, nitrate reduction, the Krebs carboxylic acid cycle and cellular protection (Okajima et al. 1975; Mengel and Kirkby 1978; Clarkson and Hanson 1980). Only a small proportion (about 10–20%) of the iron in a plant is physiologically active, and so total content is therefore not a very reliable indicator of sufficiency. For example, tissue analyses often reveal that Fe contents of chlorotic leaves are higher than healthy leaves, or even than leaves damaged by Fe toxicity, because of accumulation of Fe in the veins. Various reasons have been postulated, for instance high phosphate or bicarbonate concentrations and high pH, for the inactivation of Fe in the veins.

8.2.1 Iron Deficiency

Iron deficiency is a serious disorder in rice on neutral and alkaline aerobic soils; the severity increases with the pH of the soil. Iron deficiency limits growth even on dryland acid soils (IRRI 1963) since the Fe requirement of rice is greater than that of other plants (Ponnamperuma 1975; Agarwala and Sharma 1979). Atmospheric oxygen is supplied to the roots via the aerenchyma as an adaptation to growth in flooded soils (Armstrong 1967, 1969), which contributes to the oxidation of the rhizosphere of upland as well as of wetland rice. In wetland rice Fe deficiency occurs in alkaline soils, soils low in organic matter and in vertisols together with zinc deficiency (Ponnamperuma and Lantin 1985). Fe deficiency is also a problem in peat soils, especially if drained, low in iron and high in pH.

Young rice plants are very susceptible to iron deficiency because of their small root mass and because they secrete only small amounts of the phytosiderophore, deoxymugineic acid, which is rapidly degraded (Mori et al. 1991).

The initial symptoms of Fe deficiency in rice are yellowing or chlorosis of the interveinal areas of the emerging leaf. Later, the entire leaf turns yellow, finally white. If the deficiency is severe, the entire plant becomes chlorotic and dies. Since Fe is so immobile within the plant, Fe fertiliser (foliar or soil addition) has to be repeatedly applied as new leaves emerge. As a result, Fe deficiency is the most difficult of the nutrient deficiencies to correct.

The symptoms of Fe deficiency result from inhibited development and function of chloroplasts. The activities of Fe-related enzymes decline when the amount of iron is insufficient. Shortage of Fe restricts chlorophyll formation and function (Okajima et al. 1975; Agarwala et al. 1986). Fe deficiency results in a decrease in dry matter production, in chlorophyll content and in specific activities of enzymes such as catalase, succinic dehydrogenase and aconitase (Agarwala et al. 1986).

The problem of iron deficiency is not so much the insufficiency as the availability of iron in the soil. Under aerobic conditions iron availability is governed by the solubility of iron oxide III hydrates. The solubility of iron oxides is so low that iron released into the soil solution may not exceed the theoretical 0.001 ppm (Ponnamperuma 1972). The presence of metastable ferrosic hydroxides may explain Fe deficiency in some, but not all soils (Lindsay 1991).

The solubility of Fe is pH, Eh and temperature-dependent (Lindsay 1974); for example the concentration of Fe decreases a hundredfold for each unit pH increase (Ponnamperuma 1972). Fe deficiency is most common as a mineral stress of upland rice in neutral, calcareous and alkaline soils having a pH > 6.5. High bicarbonate ion concentrations in soils and irrigation waters aggravate Fe deficiency (Okajima et al. 1975). Insufficient O_2 supply and accumulation of bicarbonate in the rhizosphere, which hamper root respiration, are often the primary cause for reduced uptake and translocation of Fe. Chaney (1984) pointed out that bicarbonate induces Fe chlorosis in dicotyledonous, but not in monocotyledonous plants: and bicarbonate hinders the translocation of Fe out of the vascular bundles in grapevines (Mengel and Bübl 1983).

In calcareous and in alkaline soils which are low in organic matter, Fe deficiency may occur even if the soil is waterlogged. In these soils the pH remains above 7 and hence the activity of Fe is low. At high pH, Fe is readily oxidised, leading to the formation of insoluble $Fe(OH)_3$ at the root surface and this inhibits further Fe uptake. These coatings have been identified as the mineral lepidocrocite (Bacha and Hossner 1977). Fe deficiency can be induced by adding nitrate (Agarwala et al. 1986). Sodium nitrate, calcium cyanide and ammonium nitrate all aggravate Fe deficiency by raising the pH of the rhizosphere and consequently urea, ammonium sulphate and ammonium chloride are better sources of nitrogen for rice.

High applications of phosphate may induce Fe chlorosis and aggravate Fe deficiency by inactivation of Fe uptake, translocation to the shoot and subsequent metabolism. Ferric phosphate precipitates in the growing medium and in the plant sap (Kimura 1950, as cited by Okajima et al. 1975). Chlorotic plants mostly have higher P/Fe ratios than healthy plants, but according to Mengel and Scherer (1984), the increased P content is not the cause, but a consequence of Fe deficiency. Excessive amounts of Mn, Cu, Zn, Mo, Ni and Al have been reported to cause Fe deficiency (Clark et al. 1957; Okajima et al. 1975; Fageria and Carvalho 1982). These other metals compete for absorption sites or inhibit the transport and physiological reactions of Fe.

Fe deficiency can be amended through foliar or soil application of salts such as ferrous sulphate $[Fe_2(SO_4)_3]$ or Fe chelates, but Fe fertilisation can be costly and very difficult because the low mobility of Fe in the plant can mean that repeated application is necessary. Breeding for Fe efficiency is consequently an important goal and the wide variability for tolerance to Fe deficiency, not only between plant species but also cultivars (Brown 1976), opens up the possibility to screen and breed for tolerance to soils of low Fe availability.

8.2.2 Iron Toxicity

Because of the low solubility of iron in aerobic soils, toxicity is rarely found in upland crops. In wetland rice, iron toxicity severely limits production on strongly acid soils with moderate to high amounts of organic matter and reactive iron. Soil features that are also linked to iron toxicity are low pH, low cation exchange capacity, low base status, low supply of Mn, and poor drainage (Ponnamperuma 1974; Ponnamperuma and Solivas 1982). According to Eq. (8.3), the concentration of water-soluble Fe^{2+} reaches a (toxic) level of 300 mg/l at a pE = -0.76 or Eh = -45 mV and ionic strength = 0.03 mol/l at 25 °C. At pH 5 a redox potential of only +309 mV is needed. The rate of formation of toxic levels of water-soluble Fe^{2+} depends on pH, organic matter content of the soil, the nature and content of FeIII oxide hydrates and the temperature (Ponnamperuma 1972). Low temperatures cause high and persistent water-soluble iron (Cho and Ponnamperuma 1971).

Iron toxicity is found in young acid sulphate soils (Sulfaquepts); poorly drained colluvial and alluvial sandy soils (Hydraquents, Tropaquents, Fluvaquents) in valleys receiving interflow water from adjacent acid highlands; alluvial or colluvial clayey acid Tropaquepts and Tropaquents and acid peat soils.

The plant symptoms of iron toxicity are more variable than those of iron deficiency. Small brown spots appear on the lower leaves, starting at the tips. Later, the entire leaf turns brown, purple, yellow or orange, depending on the variety. The leaves of some varieties may roll. In severe cases of Fe toxicity, the lower leaves turn brown and die. Growth and tillering are depressed and the root system is coarse, scanty and dark brown. If Fe toxicity occurs late, vegetative growth is not severely affected but the grain yield is reduced because of sterility.

Roots of plants damaged by Fe toxicity are poorly developed, black, decaying and iron-coated (Benckiser et al. 1982). The contents of Fe in the leaf are not correlated with Fe toxicity. The iron content of the leaves showing symptoms is often similar to, or even lower than, leaves of healthy plants – 500 to 5000 mg/kg at maturity.

The values reported for the concentration of iron in the soil solution that constitute toxicity range from 10–1000 mg/l (Tanaka et al. 1966). The wide range indicates the lack of specific criteria for Fe toxicity, the forms of iron considered, growth stage, variety, presence of respiration inhibitors, nutrient status and environmental factors. In acid sulphate soils, Fe toxicity generally occurs because the solubility of Fe is very high (Nhung and Ponnamperuma 1966), and occurs even if plants are adequately nourished (Moormann and van Breemen 1978). In soils with low nutrient levels or respiration inhibitors concentrations as low as 20–40 mg/l have resulted in Fe toxicity (van Breemen 1978). In sandy soils with continuous supply from interflow, iron toxicity occurred at concentrations of 40–100 mg/l (van Breemen and Moormann 1978). Under acid conditions, SO_4^{2-} and Cl^- appear to cause an excessive uptake of Fe^{2+} (van Mensvoort et al. 1985).

Fe toxicity is often associated with a deficiency of P, K, Zn, Ca or Mg, strongly imbalanced nutrient solutions, or the presence of H_2S, rather than with high levels of active Fe (Ota and Yamada 1962; Mulleriyama 1966; Tanaka and Yoshida 1970; Trolldenier 1977; Ottow et al. 1982). High concentrations of total sulphides inhibit root respiration (Tanaka et al. 1968) and free sulphides destroy the ability of the roots to protect the plants from excess uptake of iron, which renders the plant vulnerable to iron toxicity (Park and Tanaka 1968). The destruction of root oxidising power is aggravated by low K (Trolldenier 1977). Moore and Patrick (1989a) suggest that uptake of divalent cations occurs at the expense of monovalent cations due to the increased production of divalent-cation carrier. In solutions dominated by Fe^{2+} this mechanism could result in K deficiency. Iron toxicity has been interpreted as a multiple nutritional stress due to insufficient nutrition of K, P, Ca and/or Mg (Howeler 1973; Benckiser et al. 1984).

Benckiser et al. (1982) reported that under conditions of low P, K, Ca and Mg availability, rice exudes metabolites, which enhances the reductive dissolution of Fe resulting in a breakdown of the iron-oxidising mechanism. Rice has a strong root oxidising power by which excess iron is oxidised and excluded from uptake (Okajima 1964; Tanaka and Navasero 1966). As well as the diffusion of atmospheric O_2 via the aerenchyma into the rhizosphere, an enzymatic root oxidation system also exists (Matsunaka 1960; Ando 1983).

Bonde (1990) suggests that the primary cause of Fe toxicity in rice is an Fe-induced excess production of oxygen radicals that surpasses the plant's capacity for control of these radicals, leading to a negative feedback system. Plants respond to various stresses with increased production of oxygen radicals (Elstner et al. 1988). Iron increases not only the rate of production but is likely to contribute highly reactive ferryl radicals (Halliwell and Gutteridge 1984). The formation of radicals is not only catalysed by Fe-ions but by various Fe-chelates (Fridovich 1986; Thompson et al. 1987), although Fe-EDTA chelates do *not* act as catalysts (Vianello et al. 1987).

Oxygen radicals, in particular, cause the autocatalysis of lipids in cell membranes (Halliwell and Gutteridge 1984; Cakmak 1988). The peroxidation of lipids takes place only when both Fe(II) and Fe(III) ions are present (Miller and Aust 1983). Detoxication of oxygen radicals is mainly brought about by the enzymes superoxide dismutase (SOD), peroxidase (POD) and catalase (Foster and Hess 1980; Matters and Scandalios 1986). SOD is mainly found in chloroplasts, mitochondria and cytosol; POD in chloroplasts, cell walls and vacuoles; and catalase in peroxisomes of green tissue (Cassab and Varner 1988). The dismutation of the oxygen radicals results in the formation of H_2O_2, which is also harmful. Catalase and/or peroxidase catalyse the final breakdown to water and oxygen.

The activities of SOD and POD in root extracts of IR9764-45-2, a rice line tolerant to Fe toxicity, were twice that of IR64, a susceptible cultivar (Bode 1990): increasing Fe concentration in the medium increased POD activity more in the Fe-tolerant rice. Catalase activity was slightly higher in the susceptible cultivar and did not respond to increased iron concentrations. To what extent deficiencies of other nutrients or high solar radiation that induce or aggravate iron toxicity influence the production and control of oxygen radicals has still to be elucidated.

Fe toxicity, as most other nutrient toxicities, is often very difficult to alleviate by management practices. Silica alleviates Fe toxicity by promoting Fe^{2+} oxidation by the roots, resulting in the deposition of iron oxides on the root surface (Horiguchi 1988). Silica-treated rice plants had lower Fe contents in the shoots and higher Fe contents in the roots than untreated plants (Okuda and Takahashi 1965). Soil reclamation can be carried out by means of liming, drainage, prolonged submergence, organic matter amendments and fertilisation with N, P, K and Zn. If the Fe/Mn ratio is very high, MnO_2 application may mitigate Fe toxicity (Ponnamperuma and Solivas 1982; Neue and Singh 1984). The prospect of breeding rice cultivars for tolerance to Fe toxicity is promising because of the wide variability that exists between cultivars (Neue 1991).

8.3 Manganese

Manganese is widely distributed in primary minerals, combined with O_2, CO_3^{2-} and SiO_2. Manganese exists in several valence states: Mn III and Mn IV oxides are the most important and abundant in aerobic soils. On flooding, the hydrous oxides are reduced to Mn^{2+}. At the pH, pE and pCO_2 values typical of reduced soils, the stable solid phases controlling Mn solubility are oxides (Mn_3O_4) and rhodochrosite ($MnCO_3$) (Ponnamperuma et al. 1969; Pasricha and Ponnamperuma 1976). The increase in concentration of water-soluble Mn^{2+} on submergence and subsequent decrease conforms to the equation (Ponnamperuma 1972):

$$pH + 0.5 \log Mn^{2+} + 0.5 \log pCO_2 = 4.4. \quad (8.4)$$

In soils with high Fe:Mn ratios, the apparent solubility of Mn can be reduced to less than the solubility of rhodochrosite because of co-precipitation with Fe (Neue 1988). In acid soils, cation exchange is the dominant mechanism governing Mn^{2+} activities (Moore and Patrick 1989b).

Plants absorb and translocate manganese mainly as Mn^{2+}. The mobility of Mn in the plant is low, but higher than that of Ca, B, Cu or Fe. The mobility of Mn is higher in graminaceous species than in dicotyledons (Bergmann 1988). Manganese behaves chemically like the alkali earths and heavy metals and competes with, or replaces, Mg^{2+}, Zn^{2+} and Fe^{2+}.

A number of enzymes require Mn^{2+} as an activator: oxidases, peroxidases, dehydrogenases decarboxylases and kinases (Okajima et al. 1975). Because of its redox properties, Mn takes part in the control of oxidation and reduction reactions and carboxylation processes. Mn^{2+} catalyses the formation of phosphatidic acid, the first step in phospholipid synthesis which is required for the development of cell membranes. A large quantity of Mn is found in the chloroplast, where it also stabilises the chloroplast structure (Bergmann 1988): chloroplasts are the cell organelles most sensitive to Mn deficiency (Mengel and Kirkby 1978). Mn deficiency reduces the quantity of both the chlorophyll and carotene pigments and this subsequently reduces the rate of photosynthesis. $Mn^{2+/3+}$ plays an essential role in the photosynthetic evolution of oxygen and incorporation of CO_2. The photosynthetic evolution of oxygen from rice leaves was observed to be reduced when plants were grown in pots using soil in which Mn deficiency had previously been observed using wheat (Kaur and Nayyar 1986). Manganese is the most important enzyme activator in the tricaboxylic acid cycle and its deficiency results in the accumulation of citric acid. Manganese deficiency as well as toxicity hamper both nitrate reduction and protein synthesis. The $Mn^{2+/3+}$ system activates indolacetic acid-oxidase and the oxidation of auxin is enhanced at Mn concentrations of 10^{-9} to 10^{-5} mol/m^3 (Hewitt 1958). Marschner (1986) considered that Mn had a key role in the protection of the photosynthetic apparatus against the deleterious action of oxygen radicals. Manganese also activates peroxidase even more effectively than Fe and is a component of superoxide dismutase. The beneficial effect of applying Mn to plants suffering from Fe toxicity may be triggered through enhanced activation of peroxidase.

The Mn requirement of rice is high and the plant can withstand excessive amounts. Manganese uptake by rice increases on flooding (Clark et al. 1957; Chaudhry and McLean 1963) and correlates with increase in the concentration in solution (Tanaka and Navasero 1966). The uptake of Mn has also been correlated with Mn^{2+}/Fe^{2+} activity ratio (Moore and Patrick 1989b), and may be modified by other nutrients such as Fe, Zn, Ca and Mg.

Roots initially accumulate Mn^{2+} until a saturation point is reached after which the ions are transported to the shoot (Ramani and Kannan 1975). Manganese is retranslocated from old to young leaves when Mn becomes deficient but mobility is low. Roots do not have a specific capacity to immobilise Mn (Tanaka and Navasero 1966).

8.3.1 Manganese Deficiency

There are a few reports on Mn deficiency in upland rice (Tanaka and Navasero 1966). In wetland rice fields Mn deficiency is rare because the soils generally contain adequate levels of available Mn (Yoshida 1981). A concentration of < 1 mg/kg in flooded soils has been found to be a useful criterion for Mn deficiency (Randhawa et al. 1978). Mn deficiency may occur in alkaline and calcareous soils with low quantities of organic matter and reducible Mn, in "degraded" paddy soils (Mitsui 1956), in leached sandy soils (Randhawa et al. 1978) and calcareous peats (Jones et al. 1980). Rice seedlings became Mn deficient when grown on a drained Histosol when the pH was near to, or above, 7 (Snyder et al. 1990).

Manganese-deficient rice plants are stunted with a pale, greyish green interveinal chlorosis. The chlorosis spreads from the tip to the base of the leaf and the chlorotic tissue later develops reddish brown necrotic spots.

In rice there are well-established interactions between Fe and Mn (Nhung and Ponnamperuma 1966; Tanaka and Navasero 1966; Jugsujinda and Patrick 1977). Antagonism with Fe combines with low Mn content to cause Mn deficiency in iron-toxic soils (Ponnamperuma and Solivas 1982). Excess aluminium in acid upland soils depresses Mn uptake (Fageria and Carvalho 1982). High concentrations of Ca^{2+}, Mg^{2+}, Fe^{2+}, Zn^{2+} and NH_4^+ can negatively affect Mn^{2+} uptake, while NO_3^- ions have a positive effect.

Alleviation of Mn deficiency is not as difficult as Fe deficiency. On acid soils, the application of $MnSO_4$ or MnO has been successful while on neutral and alkaline soils foliar application is recommended. Mn chelates are less effective when applied to the soil because Fe, Cu and other cations easily displace Mn. Some observations suggest that screening and breeding for tolerance to Mn deficiency should be possible: for instance, Fe-efficient soybeans took up not only more Fe but also Mn, Zn, and other metals including Al (Wallace and Cha 1986).

8.3.2 Manganese Toxicity

Manganese toxicity occurs more often in the field than generally expected, especially on aerobic soils that become temporarily wet and have a pH < 6. Mn toxicity depresses yields of most dryland crops on acid soils and dryland rice is no exception (IRRI 1966; Ponnamperuma 1975). Acid rains have enhanced the severity of Mn toxicity (Ulrich 1984; Keil et al. 1986). The solubility of Mn in aerobic soils increases steeply as the pH falls below 4.5 while the solubility of Fe changes little until the pH is 2.7–3.0. A decrease in pH increases the Mn to Fe ratio leading to Mn toxicity (IRRI 1971). Mn toxicity does not induce Fe deficiency and vice versa (Agarwala et al. 1986).

Flooding reduces manganese oxides to water-soluble Mn^{2+} ions. Paddy soils typically contain 50–3000 ppm reduced Mn in the solid phase and 1–100 ppm Mn^{2+} in solution. Despite this, Mn toxicity has hardly ever been observed in wetland rice. This is attributed to the high supply of Fe in wetland soils, and the natural adaptation of rice to flooded soils (Ponnamperuma 1974; Tadano and Yoshida 1978).

Visual symptoms of manganese toxicity in rice are brown spots on older leaves and drying of the tips about 8 weeks after planting. Vegetative growth is not appreciably affected but grain yield is markedly depressed because of high sterility. The symptoms of manganese toxicity differ from those iron deficiency (Tanaka and Navasero 1966).

Foy et al. (1978) attributed Mn tolerance of plants to low absorption and translocation of excess Mn to plant tops, tolerance to high Mn levels within plant tissues, oxidation of Mn in the root zone, and a physical/chemical compartmentalisation of Mn at the root. Rice possesses a high degree of Mn exclusion. The Mn content of tissues rice plants increased by only fourfold when grown at 300 ppm Mn compared with those grown at 0.1 ppm Mn. The increase in tissue Mn concentration in similar growth conditions was 109-fold for barley and 700-fold for radish (Tanaka et al. 1975). Tolerance of Mn in rice has been attributed to the ability to oxidise Mn^{2+} in the rhizosphere to the tetravalent form which decreases Mn absorption (Okajima et al. 1975; Foy et al. 1978). In addition, rice has high internal tolerance to excess Mn (7000 ppm) in the tissue as compared to other grasses (600–1200 ppm: Vlamis and Williams 1967).

Silicon alleviates Mn toxicity by decreasing Mn uptake and increasing the internal tolerance of Mn in the tissue (Horiguchi 1988). Okuda and Takahashi (1965) found Mn oxidation greater on rice roots supplied with Si. Si decreased excessive uptake of Mn and Fe. In beans, Si did not reduce Mn uptake but increased Mn solubility and mobility in the plant, resulting in a less harmful form or distribution (Horst and Marschner 1978). Manganese toxicity in rice can be alleviated by the addition of $FeSO_4$, gypsum and farm yard manure (Perumal 1961). For upland crops, liming is recommended to a pH between 5.0 and 6.5, depending upon the texture of the soil.

8.4 Zinc

Total zinc in soils ranges between 10–300 ppm. It is present in minerals, as salts, adsorbed on exchange sites and solid surfaces or substituted for Mg in clay minerals. Zinc solubility decreases 100 times for each unit increase in pH following the equation;

$$\log Zn^{2+} = 5.80 - 2\,\mathrm{pH}. \qquad (8.5)$$

Generally, Zn is taken up as Zn^{2+}, but at high pH presumably as $Zn(OH)^{+}$. Because of low concentrations in the soil solution and low mobility, Zn is mainly taken up by direct root contact. The uptake of Zn is metabolically controlled (Giordano et al. 1974), and transport in the plant is either as the free divalent cation or bound to organic acids (Kitagishi and Obata 1986), although the specific mechanism of translocation is not known. The mobility of Zn in the plant is low, but better than that of Fe, B and Mo. Zinc accumulates in the root tissues but may be retranslocated to the shoot if needed (Loneragan 1976), and Zn is also partially translocated from old leaves to developing organs.

Zinc acts as specific activator for some enzymes and influences various other metabolic processes. Zinc resembles Mn^{2+} and Mg^{2+} ions and, like Mg, it does not undergo oxidation or reduction. The functions of Zn in the plant are based on its properties as a divalent cation with a tendency to form tetrahedral complexes. Unlike Mg^{2+}, which interacts rapidly and reversibly with enzymes, Zn^{2+} tends to be tightly bound within metalloenzymes. Zinc activates carbonic anhydrase, glutamic acid dehydrogenase, lactic acid dehydrogenase, alcohol dehydrogenase, alkaline and acid phosphatase, enolase, aldolase, CuZn superoxide dismutase and RNA polymerase.

Zinc influences nitrogen metabolism, participates in protein synthesis and promotes the synthesis of cytochrome (Mengel and Kirkby 1978) and is essential in maintaining a high chlorophyll to carotenoid ratio (Kumar et al. 1976), and both Zn deficiency and toxicity reduce chlorophyll contents. Zinc deficiency decreases RNA and ribosome content and hampers protein synthesis with the result that amino acids and amides accumulate (Kitagishi and Obata 1986). Zinc generally enhances reduction processes while Cu and Mn increase oxidation processes in the redox chain. Zinc deficiency impairs phosphorylation of glucose and carbohydrate metabolism and affects auxin metabolism positively (Skoog 1940; Singh 1981). Zinc catalyses the synthesis of tryptophan (Bergmann 1988). Zn deficiency hampers cell division in meristematic tissue reducing root and shoot growth. Though not specific, Zn deficiency, like Mn deficiency; results in decreased catalase but increased peroxidase activity. Moore and Patrick (1988) found root alcohol dehydrogenase activity was lower in Zn-deficient plants than those that were Zn-sufficient. Application of zinc and manganese significantly increased the cation exchange capacity of rice roots (Singh and Bollu 1984).

8.4.1 Zinc Deficiency

Zinc deficiency is the most widespread micronutritional disorder of food crops the world over (Lopes 1980; Ponnamperuma et al. 1981). It is more common in wetland rice than in dryland crops (Castro 1977; Randhawa et al. 1978) because Zn is more readily available in upland than in submerged soils. The concentration of water-soluble Zn decreases on flooding, and bicarbonates, organic acids and other ions interfere with zinc uptake (IRRI 1970; Mikkelsen and Kuo 1977).

Zinc deficiency in wetland rice occurs on soils with pH greater than 7.0, soils with low available Zn, or low total Zn content and on soils with high amounts of organic matter (Yoshida and Tanaka 1969; IRRI 1972, 1980; Yoshida et al. 1973; Katyal and Ponnamperuma 1975). Zn deficiency has been associated with a wide range of conditions: high bicarbonate content, a magnesium to calcium ratio in soils > 1, the use of high levels of fertilisers, intensive cropping, use of high yielding varieties, prolonged submergence, and irrigation with alkaline water. The following soils are likely to be deficient in Zn: calcareous, sodic, saline-sodic and coastal saline, sandy, peaty, very poorly drained soils, as well as soils high in available P or Si and soils derived from serpentine, and regardless of pH (Tanaka and Yoshida 1970; Forno et al. 1975a; Katyal and Ponnamperuma 1975; Lantin 1977; Castro 1977; IRRI 1979; Scharpenseel et al. 1983).

In rice, visual symptoms of Zn deficiency vary with soil, variety and growth stage. Symptoms can easily be confused with N, Mg, Mn or Fe deficiency or tungro, a virus disease. Since S deficiency is often combined with Zn deficiency, it is difficult to isolate the two (Neue and Mamaril 1985). Usually, midribs at the base of the youngest leaf of Zn deficient rices become chlorotic 2–4 weeks after sowing or transplanting. Then brown spots appear on the older leaves, the spots enlarge, coalesce, and give the leaves a brown colour. Characteristic of Zn deficiency are stunted growth and depressed tillering. In moderately Zn-deficient soils, plants may recover after 4–6 weeks, but maturity is delayed and yields of susceptible cultivars are reduced (Orticio 1979).

Unlike most nutrient elements, the concentration of Zn decreases when the soil is submerged, despite desorption from FeIII and MnIV oxyhydroxides. In acid soils, much of the decrease in Zn availability upon flooding can be explained by the pH increase (Trierweiler and Lindsay 1969) and precipitation of $Zn(OH)_2$ (Lindsay 1972). The decrease in concentration of Zn in sodic and calcareous soils upon flooding, in spite of the decrease in pH, may be due to the precipitation of ZnS (Ponnamperuma 1975). Zn^{2+} activity cannot be depressed significantly by the formation of ZnS when the redox is poised by Fe reactions. If the partial pressure of CO_2 is maintained at 3 Pa at pH 7, pE-pH must drop below 5 to precipitate sulphides of Fe, Zn and Cu. Precipitation of sulphide requires an even lower redox level when carbonates control the solubility (Neue 1988).

In calcareous soils there are transient increases in the concentration of bicarbonate and organic acids upon submergence and these inhibit Zn uptake by rice plants (Lindsay 1979). Zinc may be strongly adsorbed on $CaCO_3$ or $MgCO_3$ (Katyal and Ponnamperuma 1975) since carbonate has a high affinity for zinc (Castro 1977). The breakdown of Fe and Mn oxides upon flooding provides surfaces with high absorption capacity for Zn (Ice et al. 1981) and the formation of $ZnFe_2O_4$ renders Zn unavailable (Sajwan and Lindsay 1986).

Organic matter inactivates soil Zn and retards its uptake by rice plants (IRRI 1971; Yoshida et al. 1973). Zinc deficiency is more acute in calcareous or alkaline soils where organic matter is high. At low pH, the lower values of metal-organic-matter stability constants make Zn more available (Katyal 1972). The highly significant correlation between available Zn and organic C in a toposequence of Fe-deficient soils disappeared when Mg and bicarbonate concentrations were set constant in partial correlations (Scharpenseel et al. 1983).

The increased availability of Ca, Mg, Cu, Fe, Mn and P upon flooding depress Zn availability and uptake (Giordano et al. 1974; Rashid et al. 1976; Tiwari and Pathak 1982; Sajwan and Lindsay 1986). Precipitates of hopeite [$Zn_3(PO_4)_2.H_2O$] and strong sorption in the form of Fe–Zn and Fe–P–Zn complexes exist in soil (Lindsay 1979; Gupta et al. 1987; Xie and MacKenzie 1989) and complexation of Zn–P-soil increases with addition of P (Xie and MacKenzie 1990). However, the major cause for P-induced Zn deficiency seems to be precipitation/adsorption in the root. Uptake and translocation from root to shoot is reduced with increasing P supply, especially if Zn availability is low to moderate. In maize, high P supply reduced translocation of Zn to the shoot even though it did not reduce the uptake of Zn (Trier and Bergmann 1974).

Iron absorption strongly inhibited Zn absorption and increased the Zn requirement of the rice cultivar IR6 compared to Basmati 370 (Chaudhry et al. 1977). In a related study, total Zn absorption and translocation in rice was decreased by increasing Fe in the culture medium (Giordano et al. 1974; Brar and Sekhon 1976); the inhibitory action of Fe upon Zn absorption was considered "non-competitive". Iron was deduced to displace Zn from citrate, disturbing the translocation of Zn within the plant. The Zn concentration in plant tissue is not always a reliable indicator for Zn deficiency: P/Zn and Fe/Zn ratios provide better discrimination (Neue 1991).

Addition of Ca and Mg reduced the absorption of ^{65}Zn by rice seedlings, and this was also considered to be non-competitive (Sadana and Takkar 1983). A large proportion of the Zn which is absorbed remains in the roots with only 5% translocated to the shoot. Translocation of Zn in rice seedling from roots to shoots increased upon application of Mn. Zinc and Cu mutually inhibit the uptake of each other, suggesting that both may be absorbed through the same mechanism or carrier sites (Bowen 1987). Cadmium uptake increases in Zn deficient rice (Honna and Hirata 1978), but more Zn is translocated to aerial parts than Cd (Dabin et al. 1978). Zinc addition to waterlogged soils depressed the availability of Cu, Fe and P and increased Mn (extractable by DTPA) and decreased both uptake and translocation of Cu, Fe and P (Haldar and Mandal 1981).

Amongst rice cultivars, differences in susceptibility to Zn deficiency are widely recognized (Shim and Vose 1965; IRRI 1970; Katyal 1972; Yoshida et al. 1973; Forno et al. 1975a, b); Zn-efficient rices have been selected at IRRI since 1971. The differential response to Zn deficiency has been attributed to differences in either uptake from the medium or in the ability to utilise Zn from a deficient medium. The cultivar M101 thrived while the cultivar IR26 readily developed Zn deficiency symptoms in Zn-deficient hydroponic solution. Uptake rates varied between the two cultivars (Bowen 1986): roots of M101 have a twofold greater affinity for Zn than those of IR26.

Some rice cultivars appear to have higher Zn requirements than others. Translocation of Fe, Mg and P to the shoots and the absorption of Cu decreased in IR34 when Zn became deficient (Cayton et al. 1985). The tolerance of this cultivar to Zn deficiency was attributed to a lower Zn requirement, to highly efficient translocation of Zn, and to the ability to maintain lower Fe/Zn, Cu/Zn, Mg/Zn and P/Zn ratios in the shoot than more susceptible cultivars. Tolerance to Zn deficiency is significantly correlated with tolerance to Fe toxicity in rice under high solar radiation (Neue et al. 1990; Neue 1991). Fe strongly inhibits Zn absorption and increases internal Zn requirement of susceptible rices (Sakal 1980).

Giordano and Mortvedt (1974) observed that the early maturing rice cultivar Bluebell was less tolerant to low levels of Zn than the later maturing cultivar Calrose. Generally, early maturing cultivars may be more susceptible because of the early high demand for Zn before root development is sufficient to accumulate adequate amounts (IRRI 1971).

Zn deficiency is amended by soil or foliar application of $ZnSO_4$ or Zn chelates or root dipping of rice seedlings in 2% ZnO solution. Thorough soil aeration (drying) between rice crops often alleviates Zn deficiency. On moderately Zn-deficient soils, grain yields of Zn-efficient cultivars may decrease if Zn is applied because of Zn toxicity (Cayton et al. 1985). It is, therefore, recommended to apply Zn fertiliser on moderately deficient soils only after Zn deficiency occurs.

8.5 Copper

Copper is found in igneous and sedimentary rocks. It is occluded and precipitated in soil oxides. Copper minerals in soils are complex sulphides (chalcopyrite, $CuFeS_2$; bornite, $CuFeS_4$), hydroxycarbonates and silicates ($CuSiO_3.2H_2O$) (Randhawa et al. 1978). Copper is present in the divalent form which is strongly bound to humic and fulvic acids and held tightly on inorganic exchange sites with soil pH, having a large effect on specific adsorption. Copper organic complexes play an important role in regulating the mobility and availability of Cu (Mengel and Kirkby 1978). Because of its generally low mobility, copper is taken up as Cu^{2+} ions by root interception and probably also as a chelate (Bergmann 1988).

Copper is not involved in oxidation–reduction reactions in flooded soils but its behaviour is influenced by flooding. Like Zn, Cu is released into the soil solution on submergence (IRRI 1970, 1973) but may decrease in concentration due to the precipitation, not of CuS or Cu_2S as earlier believed, but of cuprous ferrite ($Cu_2Fe_2O_4$) (Lindsay 1979). The solubility of soil Cu decreases with increasing pH and increasing organic matter content (Lindsay 1972; Ponnamperuma et al. 1981).

Lindsay (1979) described the activity-pH relationship for copper as:

$$\text{Soil-Cu} + 2H^+ = \text{Soil} + Cu^{2+} \tag{8.6}$$

$$\log Cu^{2+} = 2.8 - 2\text{ pH}. \tag{8.7}$$

The activity of Cu^{2+} ions maintained by soil Cu is about 10^{-3} times that of Zn ions in equilibrium with soil Zn. The strong complexing of copper by soil organic matter in soil solution may explain why copper deficiency is not as prevalent as zinc deficiency (Randhawa et al. 1978).

Plants take up only small amounts of Cu compared to Fe, Mn, Zn and B. In the plant, Cu is translocated as soluble Cu–N chelates (Bergmann 1988). The mobility of Cu in plants is very low, and roots have higher Cu contents than shoots even if the plants are Cu-deficient (Kamprath and Foy 1971). The small ionic diameter of Cu^{2+}, the ability to form very stable organic complexes and its electron transfer properties (Cu^+/Cu^{2+}) explain the metabolic importance of Cu either as component of, or associated with, enzymes. Close relationships exist between Cu and Mo, Mn and Fe. Copper is a regulatory factor of enzyme actions as effector, stabiliser or inhibitor (Okajima et al. 1975): Cu is regarded as the oxidation catalyst. Bound to ascorbic acid oxidase, for example, oxidation rates are 1000 times faster (Bergmann 1988) than in the absence of Cu. Its role in nitrogen and protein metabolism explains why the Cu/N ratio of young leaves is a good indicator of the Cu status of the plant. Copper-containing enzymes have a key role in photosynthesis, respiration and oxidation/reduction reactions and Cu proteins are involved in lignification, anabolic metabolism, cellular defence mechanisms and hormone metabolism. Copper is a prosthetic group in plastocyanin, superoxide dismutase, amino oxidase and cytochrome oxidase. Copper is involved in carbohydrate metabolism, nitrogen metabolism, protein metabolism and reproduction, especially pollen viability (Graham 1975; Mengel and Kirkby 1978; Alloway and Tills 1984). Toxic concentrations of Cu^{2+} inhibit the activity of enzymes such as glucosidase, invertase, catalase and urease (Bergmann 1988).

8.5.1 Copper Deficiency

Because Cu behaves chemically like Zn in submerged soils, Cu deficiency is likely to occur on Zn-deficient soils (Ponnamperuma et al. 1981). Copper deficiency occurs in soils with low total Cu and/or where Cu is in an unavailable form. Low total Cu can be due to mineralogy, soil type and high leaching (Alloway and Tills 1984) and typical Cu-deficient soils include sandy textured soils, ferrallitic soils and calcareous soils. Organic matter is the primary constituent absorbing Cu, and this explains the widespread occurrence of Cu deficiency in peat soils (Driessen 1978) and latent Cu deficiency in mineral soils with high organic matter.

High P, K and N supply can aggravate Cu deficiency: high N supply may even induce Cu deficiency (Cheshire et al. 1982). Zinc and Cu interact at absorption sites and during translocation from root to shoot (Haldar and Mandal 1981): absorption of Cu from solutions is strongly inhibited by Zn (Kausar et al. 1976; Haldar and Mandal 1981) and vice versa (Giordano et al. 1974). According to Bowen (1987), Cu and Zn are absorbed by the same uptake mechanism which is different from that for B and Mn.

Visual symptoms of Cu deficiency in rice are not well established. Plants raised in sand culture with less than 1 µg/l Cu showed chlorosis of the young leaves at 4 to 5 weeks after sowing. Tillering was depressed and yield could be greatly reduced because of sterility. In general, visible symptoms of copper deficiency are bluish-green leaves that become chlorotic near the tips. Chlorosis develops downward along both sides of the midrib, followed by dark brown necrosis of the tips. New leaves fail to unroll and maintain a needle-like appearance of the entire leaf, or occasionally in half the leaf, with the basal portion developing normally (Alloway and Tills 1984).

Copper deficiency can be amended through repeated soil or foliar application of $CuSO_4$ or Cu chelates.

8.6 Boron

Boron is present at very low concentration in the earth's crust (< 10 ppm). It is associated with highly insoluble tourmaline minerals from which it is slowly released (Mitchell 1964). Total B concentrations in soils range from 1–100 ppm, the highest B contents being found in soils of marine sediments and volcanic rocks (Krauskopf 1972; Kovda et al. 1973).

In the soil, B is present in minerals, adsorbed on clays and hydrous Fe/Al oxides, complexed with organic matter and free as non-ionised H_3BO_3 and as the $B(OH)_4^-$ anion. Boron availability decreases with increase in pH, in contrast to other anions. Below pH 6, B is present in the soil solution as undissociated $B(OH)_3$. Above pH 6, it is increasingly dissociated and hydrated to $B(OH)_4^-$. Accordingly, B adsorption onto organic matter, sesquioxides and clay minerals increases with rising soil pH. Boron supply is higher in heavy textured soils where illitic clay adsorbs more B than montmorillonitic or kaolinitic clays (Keren and Bingham 1985). Release of B increases with temperature.

Boron is transported in the soil by mass flow and diffusion, therefore higher soil moisture contents and flooding generally increase B supply to plants. In acid soils, the increase of pH upon flooding may apparently decrease B supply. Boron uptake is closely related to B concentrations of the soil solution and the amount of water transpired.

Boron mobility in the plant is low and retranslocation seems highly restricted. Most B is bound as an ester to membranes and cell walls. The physiological role of B is not well defined but a multiplicity of functions have been proposed (Mengel and Kirkby 1978). The ability of B to form esters and polyhydroxyl links with organic complexes suggests similarities to the phosphate anion. Boron is not an enzyme component but seems to increase the activity of some (Bergmann 1988). Effects of boron on growth and cell development depend on the formation of complexes with reactants and products of enzymatic reactions (Dugger 1983). Boron has an essential role in the development and growth of new cells in the plant meristem, has been postulated to stabilise cell wall constituents including the plasma membrane, and to influence carbohydrate metabolism, phenol metabolism, RNA metabolism and the synthesis of sugars, starch and protein, as well as transpiration and translocation (Rains 1976; Tisdale et al. 1985). Boron is required for seed and grain production.

8.6.1 Boron Deficiency

Because of the various soil properties that control B availability, deficiency occurs especially in a wide range of upland soils. The area of B-deficient soils has been estimated to be larger than for any other micronutrients. Over-liming of soils may result in B deficiency. Boron deficiency is not common in wetland rice soils, but vast areas of highly weathered acid red soils and acid sandy rice soils in China are deficient in B (Liu Zheng and Zhu Qi-qing 1981).

In B-deficient rice plants, the tips of the emerging leaves turn white and then roll, as in Ca deficiency (Tanaka and Yoshida 1970). Plant height is reduced. A boron content of less than 20 mg/kg in young rice leaves at tillering indicates deficiency (Agarwala and Sharma 1979). B deficiency is alleviated by soil or foliar application of boric acid, borax or B-containing mixed fertiliser.

8.6.2 Boron Toxicity

Toxicity to rice in the field due to B was first observed on a soil irrigated with water with a high B content from deep wells at the IRRI Farm (Ponnamperuma 1979). Boron toxicity occurs in arid irrigated areas, volcanic areas, saline, sodic and coastal soils. Ponnamperuma and Yuan (1966) identified B toxicity as a limiting factor for rice yield on a coastal saline silt loam. Dependent upon evaporation, amount of irrigation water applied, drainage and soil type, B concentrations of $> 0.5 - 2.0$ mg/l are considered hazardous.

The first symptoms of B toxicity in rice occur as a light brown or yellowish white discoloration on tips and margins of the older leaves about 6 weeks after transplanting. As the disorder progresses, the tips and leaf margins turn yellow. Two to four weeks later, elliptical dark brown blotches appear in the discoloured areas in most rice cultivars. Finally, the entire leaf blade turns light brown and withers. Vegetative growth is not markedly depressed unless the toxicity is severe.

Rice cultivars react to excess B differently in their visible symptoms and grain yield; severity of necrotic spots is not, however, a good indicator of susceptibility. Tolerant cultivars suffered a 10–20% yield reduction when the mud extract had at least 3 mg B/l throughout the growing period (Cayton 1985).

Boron toxicity in acid soils can be amended by liming. In neutral and alkaline soils excess B has to be washed out. High soil temperatures in the tropics enhance the leaching of B.

8.7 Molybdenum

Total content of molybdenum in soils is in the range of 1–5 mg/kg. The valency of Mo ions depend strongly on the pH. Between pH 2.5 and 5.0, undissociated H_2MoO_4 is present along with $HMoO_4^-$ and MoO_4^{2-} anions: above pH 5 the MoO_4^{2-} anions predominate. In soils, molybdenum is adsorbed on positively charged sites; oxides and hydroxides of Fe are of prime importance followed by aluminium oxides, halloysites and kaolinites (Mengel and Kirkby 1978; Jones et al. 1980). Because it is present as an anion, the behaviour of molybdate in soils resembles that of phosphate. The solubility of Mo is very low and, in contrast to

B, Cu, Fe, Mn and Zn, its availability increases with the pH; it is strongly adsorbed in acid soils. There is limited information about the behaviour of Mo-containing minerals but, based on a limited number of soils, Lindsay (1979) provided the following empirical relationship:

$$\text{Soil-Mo} <\text{---}> MoO_4^{2-} + 0.8\ H^+ \qquad \log K = -12.40. \tag{8.8}$$

Flooding increases water-soluble Mo due to desorption from Fe(III) oxide hydrates following a rise in pH and reduction (Ponnamperuma 1972). At very low redox levels the precipitation of MoS_2 may lower MoO_4^{2-} activities to extremely low levels.

Molybdenum is adsorbed as molybdate (Clarkson and Hanson 1980). It competes with SO_4^{2-} in root uptake, which has been attributed to the similar size of the two anions (Stout et al. 1951). Molybdenum accumulates in roots and its mobility in plants is low. Seeds of legumes are mostly very rich in Mo.

Molybdenum is an essential constituent of flavoprotein enzymes such as aldehyde oxidase, hydrogenase, nitrogenase and nitrate reductase. Nitrate reductase, a NADH flavomolybdate protein, functions only with Mo to reduce NO_3^- to NO_2^- (Mengel 1984). Molybdenum is also essential for the synthesis of nicotinic acid (Dennis 1971): Mo enhances phosphorylation and inhibits acid phosphatase.

Molybdenum deficiency is rare in wetland rice. Rice plants grown in flooded soils need less nitrate reductase because N is absorbed as NH_3, NH_4OH, or NH_4^+ and flooding increases the concentration of water-soluble Mo. In dryland crops Mo deficiency occurs on strongly acid soils, where it is strongly adsorbed and the presence of high Al and Mn concentrations accentuate Mo deficiency. In organic soils, deficiency may occur because of retention of Mo by insoluble humic acid (Mengel and Kirkby 1978). Tang Li-Hua (1980) reported that although wetland rice shows no response to Mo fertiliser, there are large areas of Mo-deficient paddy fields on which other crops before or after rice respond to Mo fertilisers. Molybdenum-deficient soils are strongly acid in reaction or low in total Mo. Lantin (1976) obtained yield increases when Zn, Cu and Mo were applied to peat soils. The first reported positive response to Mo in wetland rice was obtained on three Histosols (IRRI 1978). Foliar symptoms of Mo deficiency in rice are paling of apical margins of the middle leaves. Molybdenum deficiency may also induce symptoms similar to N or Fe deficiency.

Molybdenum deficiency can be amended through soil or foliar application of Na and NH_4^- molybdates or Mo-superphosphate. However, liming the soil to adequate pH levels is often more effective.

References

Agarwala SC, Sharma CP (1979) Recognizing micronutrient disorders of crop plants on the basis of visible symptoms and plant analysis. Bot Dep, Lucknow University, Lucknow, India

Agarwala SC, Chatterjee C, Nautiyal N (1986) Effect of Mn supply on the physiological availability of Fe in rice plants grown in sand culture. Soil Sci Plant Nutr 32: 169–178
Alloway BJ, Tills AR (1984) Copper deficiency in world crops. Outlook Agric 13: 32–41
Amberger A, Gatser R, Ulunsch A (1982) Iron chlorosis induced by high copper and manganese supply. J Plant Nutr 5: 715–720
Ando T (1983) Nature of oxidizing power of rice roots. Plant Soil 72: 57–71
Armstrong W (1967) The oxidizing activity of roots in waterlogged soils. Physiol Plant 20: 920–926
Armstrong W (1969) Rhizosphere oxidation in rice: an analysis of intervarietal differences in oxygen flux from the roots. Physiol Plant 22: 296–303
Arnon DI, Stout PR (1939) Molybdenum as an essential element for higher plants. Plant Physiol 14: 599–602
Baba I, Harada T (1954) Physiological disease of rice plants in Japan. Ministry of Agriculture and Forestry, Tokyo
Bacha RE, Hossner LR (1977) Characteristics of coatings formed on rice roots as affected by iron and manganese additions. Soil Sci Am J 41: 931–935
Benckiser G, Ottow JCG, Santiago S, Watanabe I (1982) Physicochemical characterization of iron toxic soil in some Asian countries. IRRI Res Pap Ser 85: 1982
Benckiser G, Santiago S, Neue HU, Watanabe I, Ottow JCG (1984) Effect of fertilization on exudation, dehydrogenase activity, iron-reducing population and Fe^{2+} formation in the rhizosphere of rice (*Oryza sativa* L.) in relation to iron toxicity. Plant Soil 79: 305–316
Bergmann W (1988) Ernährungsstörungen bei Kulturpflanzen, 2 Aufl. Fischer, Stuttgart
Bode K (1990) Untersuchungen zur Eisentoleranz von Reispflanzen. MS Thesis, Institut für Allgemeine Botanik, Universität Hamburg
Bowen JE (1986) Kinetics of zinc uptake by two rice cultivars. Plant Soil 94: 99–107
Bowen JE (1987) Physiology of genotypic differences in zinc and copper uptake in rice and tomato. Plant Soil 99: 115–125
Brar MS, Sekhon GS (1976) Interaction of zinc with other micronutrient cations. I. Effect of copper on $zinc^{65}$ absorption by wheat seedlings and its translocation within the plants. Plant Soil 45: 137–143
Brown JC (1976) Genetic potentials for solving problems of soil mineral stress: iron deficiency and boron toxicity in alkaline soils. In: Wright MJ, Ferrari SA (eds) Plant adaptation to mineral stress in problem soils. Cornell Univ, Ithaca NY, pp 83–94
Brown JC (1978) Mechanism of iron uptake by plants. Plant Cell Environ 1: 249–258
Broyer TC, Carlton CM, Johnson CM, Stout PR (1954) Chlorine – a micronutrient element for higher plants. Plant Physiol 29: 526–532
Cakmak I (1988) Morphologische und physiologische Veränderungen bei Zink-Mangelpflanzen. Dissertation, Institut für Pflanzenernährung, Universität Hohenheim
Cassab GJ, Varner JE (1988) Cell wall proteins. Annu Rev Plant Physiol Plant Mol Biol 39: 321–353
Castro RU (1977) Zinc deficiency in rice: a review of research at the International Rice Research Institute. IRRI Res Pap Ser 9, International Rice Research Institute, PO Box 933, Manila, Philippines
Cayton MTC (1985) Boron toxicity in rice. IRRI Res Pap Ser 113, 10 pp
Cayton MTC, Reyes ED, Neue HU (1985) Effect of zinc fertilization on the mineral nutrition of rices differing in tolerance to zinc deficiency. Plant Soil 87: 319–327
Chaney RL (1984) Diagnostic practices to identify iron deficiency in higher plants. J Plant Nutr 7: 47–67
Chaney RL, Brown JC, Tiffin LO (1972) Obligatory reduction of ferric chelates in iron uptake by soybeans. Plant Physiol 50: 208–213
Chaudhry FM, Alam SM, Rashid A, Latif A (1977) Mechanism of differential susceptibility of two rice varieties to zinc deficiency. Plant Soil 46: 637–642
Chaudhry MS, McLean EO (1963) Comparative effects of flooded and unflooded soil conditions and nitrogen application on growth and nutrient uptake by rice plants. Agron J 55: 565–567

Cheshire MV, Bick W, de Kock PC, Inkson RH E (1982) The effect of copper and nitrogen on the amino acid composition of oat straw. Plant Soil 66: 139–147
Cho DY, Ponnamperuma FN (1971) Influence of soil temperature on the chemical kinetics of flooded soils and the growth of rice. Soil Sci 112: 184–194
Clark F, Nearpass DC, Specht AU (1957) Influence of organic addition and flooding on iron and manganese uptake by rice. Agron J 49: 586–589
Clarkson DT, Hanson JB (1980) The mineral nutrition of higher plants. Annu Rev Plant Physiol 31: 239–298
Dabin P, Marafante E, Mousny JM, Myttenaere C (1978) Absorption distribution and binding of cadmium and zinc in irrigated rice plants. Plant Soil 50: 329–341
Dennis EJ (1971) Micronutrients – a new dimension in agriculture. Publ Nation Fert Sol Assoc, Peoira, Ill, USA
Driessen PM (1978) Peat soils. In: Soils and rice. International Rice Research Institute, Manila, Philippines, pp 763–779
Dugger W (1983) Boron in plant metabolism [nutrients]. In: Encyclopedia of plant physiology, New Series, vol 15B. Inorganic plant nutrition. Springer, Berlin Heidelberg New York, pp 626–650
Elstner EF, Wagner GA, Schutz W (1988) Activated oxygen in green plants in relation to stress situations. Curr Top Plant Biochem Physiol 7: 159–187
Fageria NK, Carvalho JRP (1982) Influence of Al in nutrient solutions on chemical composition in upland rice cultivars. Plant Soil 69: 31–44
Forno DA, Asher CJ, Yoshida S (1975a) Zinc deficiency in rice. I. Soil factors associated with the deficiency. Plant Soil 42: 537–550
Forno DA, Asher CJ, Yoshida S (1975b) Zinc deficiency in rice. II. Studies on two varieties differing in susceptibility to Zn deficiency. Plant Soil 42: 551–563
Foster JG, Hess JL (1980) Responses of SOD and glutathione reductase activities in cotton leaf tissue exposed to an atmosphere enriched in oxygen. Plant Physiol 66: 482–487
Foy CD, Chaney RL, White MC (1978) The physiology of metal toxicity in plants. Annu Rev Plant Physiol 29: 511–566
Fridovich I (1986) Superoxide dismutases. Adv Enzymol 41: 35–97
Giordano PM, Mortvedt JJ (1974) Response of several rice cultivars to zinc. Agron J 66: 220–223
Giordano PM, Noggle JC, Mortvedt JJ (1974) Zinc uptake by rice as affected by metabolic inhibitors and competing cations. Plant Soil 41: 637–646
Graham RD (1975) Male sterility in wheat plants deficient in copper. Nature 254: 514–515
Gris E (1844) Nouvelles experiences sur l'action des composés ferrugineux solubles, appliqués à la végétation, et specialement au trâitement de la chlorose et de la debilité des plantes. C R Acad Sci Paris 19: 1118–1119
Gris E (1847) Addition à une precedente. Note concernant des experiences sur l'application des sels de fer à la vegetation, et specialement au traitement des plantes chlorosées, languissantes et menacées d'une mort prochaine. C R Acad Sci Paris 25: 276–278
Gupta RK, Elihout SVD, Abrol IP (1987) Effect of pH on zinc adsorption–precipitation reactions in an alkali soil. Soil Sci 143: 198–204
Haldar M, Mandal LN (1981) Effect of P and Zn on the growth and P, Zn, Cu, Fe and Mn nutrition of rice. Plant Soil: 415–425
Halliwell B, Gutteridge JMC (1984) Oxygen toxicity, oxygen radicals, transition metals and disease. Biochem J 219: 1–14
Hewitt EJ (1958) The role of mineral elements in the activity of plant enzyme systems. In: Ruhland W (ed) Handbuch Pflanzenphysiologie, BdIV. Die mineralische Ernährung der Pflanze. Springer, Berlin Heidelberg New York, pp 427–481
Honna Y, Hirata H (1978) Noticeable increase in cadmium absorption by zinc deficiency rice plants. Soil Sci Plant Nutr 24: 295–297
Horiguchi T (1988) Mechanism of manganese toxicity and tolerance of plants. IV. Effects

of silicon on alleviation of manganese toxicity of rice plants. Soil Sci Plant Nutr 34: 65–73

Horst WJ, Marschner H (1978) Einfluss von Silizium auf den Bindungszustand von Mangan im Blattgewebe von Bohnen (*Phaseolus vulgaris*). Z Pflanzenernaehr Bodenkd 141: 487–497

Howeler RH (1973) Iron induced oranging disease of rice in relation to physicochemical changes in a flooded Oxisol. Soil Sci Soc Am Proc 37: 898–903

Ice KL, Pulfind ID, Duncan HJ (1981) Influence of waterlogging and lime and organic matter addition on the distribution of trace metals in an acid soil. II. Zn and Cu. Plant Soil 59: 327–333

IRRI (1963) Annu Rep 1962. International Rice Research Institute, Manila, Philippines

IRRI (1966) Annu Rep 1965. International Rice Research Institute, Manila, Philippines

IRRI (1970) Annu Rep 1969. International Rice Research Institute, Manila, Philippines

IRRI (1971) Annu Rep 1970. International Rice Research Institute, Manila, Philippines

IRRI (1972) Annu Rep 1971. International Rice Research Institute, Manila, Philippines

IRRI (1973) Annu Rep 1972. International Rice Research Institute, Manila, Philippines

IRRI (1978) Annu Rep 1977. International Rice Research Institute, Manila, Philippines

IRRI (1979) Annu Rep 1978. International Rice Research Institute, Manila, Philippines

IRRI (1980) Annu Rep 1979. International Rice Research Institute, Manila, Philippines

IRRI (1989) IRRI Toward 2000 and beyond. International Rice Research Institute, Manila, Philippines

IRRI (1991) World rice statistics 1990. International Rice Research Institute, Manila, Philippines

Jones US, Katyal JC, Mamaril CP, Park CS (1980) Wetland rice-nutrient deficiencies other than nitrogen. In: Rice research strategies for the future. International Rice Research Institute, Manila, Philippines, pp 327–378

Jugsujinda A, Patrick WH Jr (1977) Growth and nutrient uptake by rice under controlled oxidation–reduction and pH conditions in flooded soil. Agron J 69: 705–710

Kamprath EJ, Foy CD (1971) Lime-fertilizer-plant interactions in acid soils. In: Ulsor RA, Army TJ, Hamway JJ, Kilmer VJ (eds) Fertilizer technology and use, 2nd edn. Soil Sci Soc Am, Madison, WI, pp 105–151

Katyal JC (1972) A study of Zn equilibria in flooded soils and amelioration of Zn-deficient soils of Agusan del Norte. Terminal Report. International Rice Research Institute, Manila, Philippines

Katyal JC, Ponnamperuma FN (1975) Zinc deficiency: a widespread nutritional disorder of rice in Agusan del Norte. Philipp Agric J 58: 79–89

Kaur NP, Nayyar VK (1986) Some physiological studies on rice grown on manganese-deficient soil. IRRI Newsl 11(1): 29–30

Kausar MA, Chaudhry FM, Rashid A, Latif A, Alam SM (1976) Micronutrient availability to cereals from calcareous soils. I. Comparative Zn and Cu deficiency and their mutual interaction in rice and wheat. Plant Soil 45: 397–410

Keil P, Hecht-Buchholz Ch, Ortmann U (1986) Zum Einfluss von erhöhtem Manganangebot auf Fichtensämlinge. Allg Forstztg 41: 855–858

Keren R, Bingham FT (1985) Boron in water, soils and plants. Adv Soil Sci 1: 229–276

Kitagishi K, Obata H (1986) Effects of zinc deficiency on the nitrogen metabolism of meristematic tissues of rice plants with reference to protein synthesis. Soil Sci Plant Nutr 32: 397–405

Kovda VA, van der Berg C, Hogan RG (eds) (1973) International source book. Irrigation, drainage and salinity. FAO/UNESCO, Hutchinson, London, pp 206–290

Krauskopf KG (1972) Geochemistry of micronutrients. In: Mortvedt JJ, Giordano PM, Lindsay WL (eds) Micronutrients in agriculture. Soil Sci Soc Am Madison, WI, pp 7–36

Kumar B, Gangwar MS, Rathore VS (1976) Effect of dimethyl sulfoxide (DMSO) on zinc availability (L-value), growth and metabolic activities of rice plants. Plant Soil 45: 235–246

Lantin RS (1977) The factors limiting growth of rice on peat soil. Grains J 2: 17–19

Lindsay WL (1972) Inorganic phase equilibria of micronutrients in soils. In: Mortvedt JJ, Giordano PM, Lindsay WL (eds) Micronutrients in agriculture. Soil Sci Soc Am, Madison, WI, p 41
Lindsay WL (1974) Role of chelation in micronutrient availability. In: Carson EW (ed) The plant root and its environment. University of Virginia Press, Charlottesville, pp 507–522
Lindsay WL (1979) Chemical equilibria in soils. Wiley, New York
Lindsay WL (1991) Iron oxide solubilization by organic matter and its effect in iron availability. Plant Soil 130: 27–34
Liu Zheng, Zhu Qi-qing (1981) The status of micronutrient in relation to crop production in paddy soils in China. 1. Boron. In: Proc Symp paddy soils, Inst of Soil Science, Academia Silica. Science Press, Beijing, pp 825–831
Loneragan JF (1976) Plant efficiencies in the use of B, Co, Cu, Mn and Zn. In: Proc Worksh Plant Adaptation to Mineral Stresses in Problem Soils. Cornell University, Ithaca, NY, pp 193–203
Lopes AS (1980) Micronutrients in soils of the tropics as constraints to food production. In: Soil-related constraints to food production in the tropics. International Rice Research Institute, Manila, Philippines, pp 277–298
Marschner H (1986) Mineral nutrition of higher plants. Academic Press, London
Marschner H, Römheld V, Kissel M (1986) Different strategies in higher plants in mobilization and uptake of iron. J Plant Nutr 9: 695–713
Marschner H, Römheld V, Kissel M (1987) Localization of phytosiderophore release and of iron uptake along intact barley roots. Physiol Plant 71: 157–162
Marschner H, Römheld V, Zhang FS (1990) Mobilization of mineral nutrients in the rhizosphere by root exudates. Trans 14th Int Congr Soil Sci II: 159–163
Matsunaka S (1960) Studies on the respiratory enzyme system of plants. I. enzymatic oxidation of α-naphthylamine in rice roots. J Biochem 47: 820–829
Matters G, Scandalios JG (1986) Effect of free radical generating herbicide paraquat on the expression of the superoxide dismutase genes in maize. Biochim Biophys Acta 882: 29–38
Mengel K (1984) Ernährung und Stoffwechsel der Pflanze, 6. Aufl. Fischer, Jena
Mengel K, Bübl W (1983) Verteilung von Eisen in Blättern von Weinreben mit HCO_3^--induzierter Fe-Chlorose. Z Pflanzenernaehr Bodenkd 146: 560–571
Mengel K, Kirkby EA (1978) Principles of plant nutrition. International Potash Institute, Worklaufen, Switzerland
Mengel K, Scherer HW (1984) Iron distribution in vine leaves with HCO_3^- induced chlorosis. J Plant Nutr 7: 715–724
Mikkelsen DS, Kuo S (1977) Zinc fertilization and behavior in flooded soils. Spec Publ 5. Commonwealth Bureau of Soils, Harpenden UK
Miller DM, Aust SD (1983) Studies of ascorbate dependent iron catalyzed lipid peroxidation. Arch Biochem Biophys 271: 113–119
Mitchell RL (1964) Trace elements in soils. In: Bear FR (ed) Chemistry of the soil, 2nd edn. Reinhold, New York, pp 320–368
Mitsui S (1956) Inorganic nutrition, fertilization, and soil amelioration for lowland rice. Yokendo, Tokyo
Moore PA, Patrick WH Jr (1988) Effect of zinc deficiency on alcohol dehydrogenase activity and nutrient uptake in rice. Agron J 80: 882–885
Moore PA, Patrick WH Jr (1989a) Iron availability and uptake in acid sulphate soils. Soil Sci Soc Am J 53: 471–476
Moore PA, Patrick WH Jr (1989b) Manganese availability and uptake by rice in acid sulphate soils. Soil Sci Soc Am J 53: 104–109
Moormann FR, van Breemen N (1978) Rice: soil, water, land. International Rice Research Institute, Manila, Philippines
Mori S, Nishizawa N, Hayashi H, Chino M, Yoshimura E, Ishikara J (1991) Why are young rice plants highly susceptible to iron deficiency? Plant Soil 130: 143–156
Mulleriyama RP (1966) Some factors influencing bronzing – a physiological disease of rice in Ceylon. MS Thesis, University of the Philippines, Los Baños

Neue HU (1988) Holistic view of chemistry of flooded soils. In: Panichapon GS, Wada H (Scientific eds) Elliott CR, Leslie RN (publ. eds) Proc 1st Int Symp Paddy Soil Fertility. IBSRAM, Bangkok, pp 21–53
Neue HU (1991) Adverse soil tolerance of rice: mechanisms and screening techniques. In: Deturck P, Ponnamperuma FN (eds) Rice Production on acid soils in the tropics. Institute of Fundamental Studies, Kandy, Sri Lanka, pp 243–250
Neue HU, Bloom PR (1989) Nutrient kinetics and availability in flooded soils. In: Progress in irrigated rice research. International Rice Research Institute, Manila, Philippines, pp 173–190
Neue HU, Mamaril CP (1985) Zinc, sulfur and other micronutrients in wetland soils. In: Wetland soils: characterization, classification and utilization. International Rice Research Institute, Manila, Philippines, pp 307–320
Neue HU, Singh VP (1984) Management of wetland rice and fishponds on problem soils in the tropics. In: Petersen JB (ed) Ecology and management of problem soils in Asia. FFTC Book Ser 27, Taipei, pp 352–366
Neue HU, Lantin RS, Cayton MTC, Autor NU (1990) Screening of rices for adverse soil tolerance. In: El Bassam N, Dambroth M, Loughman BC (eds) Genetic aspects of plant nutrition. Kluwer Academic, Dordrecht, pp 523–531
Nhung MT, Ponnamperuma FN (1966) Effect of calcium carbonate, manganese dioxide, ferric hydroxide, and prolonged flooding on chemical and electrochemical changes and growth of rice in a flooded acid sulphate soil. Soil Sci 102: 29–41
Okajima H (1964) Environmental factors and nutrient uptake. In: Proc Symp mineral nutrition of the rice plant. Hopkins, Baltimore, pp 63–73
Okajima H, Uritani I, Kun-huang H (1975) The significance of minor elements on plant physiology. Food and Fertilizer Technology Center, Taipei, Taiwan, Rep China
Okuda A, Takahashi E (1965) The role of silicon. In: The mineral nutrition of the rice plant. IRRI. Hopkins, Baltimore MD, pp 123–146
Orticio MR (1979) Zinc deficiency: a widespread nutritional disorder of rice in the Philippines. Saturday Seminar, International Rice Research Institute, Manila, Philippines
Ota Y, Yamada N (1962) Physiological study on bronzing of rice plant in Ceylon (preliminary report). Proc Crop Sci Soc Jpn 31: 90–97
Ottow JCG, Benckiser G, Watanabe I (1982) Iron toxicity of rice as a multiple nutritional soil stress. Trop Agric Res Ser 15: 167–179
Park YD, Tanaka A (1968) Studies of the rice plant man "akiochi" soil in Korea. Soil Sci Plant Nutr 14: 27–34
Pasricha NS, Ponnamperuma FN (1976) Na^+–(Ca^{2+} + Mg^{2+}) exchange equlibria under submerged soil conditions. Soil Sci 123: 220–223
Perumal S (1961) Leaf-tip-drying disease on rice (*Oryza sativa*). Soil Sci 91: 218–221
Ponnamperuma FN (1972) The chemistry of submerged soils. Adv Agron 24: 29–95
Ponnamperuma FN (1974) Problem rice soils. In: Proc Int Rice Research Conf, April 22–25, 1974. International Rice Research Institute, Philippines, p 11
Ponnamperuma FN (1975) Growth-limiting factors of aerobic soils. In: Major research in upland rice. International Rice Research Institute, Manila, Philippines, pp 40–43
Ponnamperuma FN (1979) Soil Problems in the IRRI Farm. Paper presented at a Thursday seminar, 8 November 1979. International Rice Research Institute, Manila, Philippines
Ponnamperuma FN, Lantin RS (1985) Diagnosis and amelioration of nutritional disorders of rice. In: Int Rice Res Conf, International Rice Research Institute, Manila, Philippines
Ponnamperuma FN, Solivas JL (1982) Field amelioration of an acid sulphate soil with manganese dioxide and lime. In: Dost H, van Breemen N (eds) Proc Bangkok Symp acid sulphate soils. International Institute for Land Reclamation and Improvement, Wageningen, pp 213–222
Ponnamperuma FN, Yuan WL (1966) Toxicity of boron to rice. Nature 211: 780–781
Ponnamperuma FN, Loy TA, Tianco EM (1969) Redox equilibria in flooded soils. II. The manganese oxide systems. Soil Sci 108: 48–57

Ponnamperuma FN, Cayton MTC, Lantin RS (1981) Dilute hydrochloric acid as an extractant for available Zn, Cu and B in rice. Plant Soil 61: 297–310
Rains DV (1976) Mineral metabolism. In: Bonner J, Varner JE (eds) Plant Biochemistry. Academic Press, New York, pp 561–597
Ramani S, Kannan S (1975) Manganese absorption and transport in rice. Physiol Plant 33: 133–137
Randhawa NS, Sinha MA, Takkar PN (1978) Micronutrients. In: Soils and rice. International Rice Research Institute, Manila, Philippines, pp 581–604
Rashid A, Chaudry FM, Sharif M (1976) Micronutrient availability to cereals from calcareous soils. III. Zn absorption by rice and its inhibitions by important ions, in submerged soils. Plant Soil 45: 613–623
Römheld U, Marschner H (1986) Mobilization of iron in the rhizosphere of different plant species. In: Tinder B, Läuchli A (eds) Advances in Plant Nutrition, vol 2. Praeger Scientific, New York, pp 155–205
Sadana US, Takkar PN (1983) Effect of calcium and maganesium on ^{65}Zn absorption and translocation in rice seedlings. J Plant Nutr 6: 705–715
Sajwan KS, Lindsay WL (1986) Effects of redox on zinc deficiency in paddy rice. Soil Sci Soc Am J 50: 1264–1269
Sakal R (1980) Iron and zinc nutrition of rice. J Indian Soc Soil Sci 28: 547–549
Scharpenseel HW, Eichwald E, Haupenthal Ch, Neue HU (1983) Zinc deficiency in a soil toposequence grown to rice at Tiaong, Quezon Province, Philippines. Catena 10: 115–132
Schwab AP, Lindsay WL (1983) The effect of redox on the solubility and availability of iron. Soil Sci Soc Am J 47: 201–205
Shim SC, Vose PB (1965) Varietal difference in the kinetics of iron uptake. J Exp Bot 16: 216–232
Singh B, Bollu RP (1984) Effect of chelated and inorganic zinc and manganese on cation exchange capacity of rice roots. Oryza 21: 167–169
Singh M (1981) Effect of zinc, P and N on tryptophan concentrations in rice grains grown on limed and unlimed soils. Plant Soil 62: 305–308
Skoog F (1940) Relationships between zinc and auxin in the growth of higher plants. Am J Bot 27: 935–951
Snyder GH, Jones DB, Coale FJ (1990) Occurrence and correction of manganese deficiency in Histosol grown rice. Soil Sci Soc Am J 54: 1634–1638
Sommer AL (1931) Copper as an essential element for plant growth. Plant Physiol 6: 339–345
Stout PR, Meagber WR, Pearson GA, Johnson CM (1951) Molybdenum nutrition of crop plants. I. The influence of phosphate and sulphate on the absorption of molybdenum from soils and solution cultures. Plant Soil 3: 51–87
Sugiura Y, Tanaka H, Mino Y, Yoshida T, Ota N, Inove M,. Nomoto K, Yoshiota H, Takemata T (1981) Structure, properties and transport mechanism of iron (III) complex of mugineic acid, a possible phytosiderophere. J Am Chem Soc 103: 6979–6982
Tadano T, Yoshida S (1978) Chemical changes in submerged soils and their effect on rice growth. In: Soils and rice. International Rice Research Institute, Manila, Philippines, pp 399–420
Takagi S (1976) Naturally occurring chelating compounds in oat and rice root washings. I. Activity measurement and preliminary characterization. Soil Sci Plant Nutr 22: 423–433
Tanaka A, Navasero SA (1966) Interaction between iron and manganese in the rice plant. Soil Sci Plant Nutr 12: 29–33
Tanaka A, Yoshida S (1970) Nutritional disorder of the rice plant in Asia. Int Rice Res Inst Tech Bull 10. International Rice Research Institute, Manila, Philippines
Tanaka A, Loe R, Navasero SA (1966) Some mechanisms involved in the development of iron toxicity symptoms in the rice plant. Soil Sci Plant Nutr 12: 32–38
Tanaka A, Mulleriyama RP, Yasu T (1968) Possibility of hydrogen sulfide-induced iron toxicity of the rice plant. Soil Sci Plant Nutr 14: 1–6

Tanaka A, Tadano T, Fujiyama H (1975) Comparison of adaptability to heavy metals among crop plants. (1) Adaptability to manganese studies on the comparative plant nutrition. J Sci Soil Manure 46: 425–430 (in Japanese)
Tang Li-Hua (1980) The status of micronutrients in relation to crop production in paddy soil of China. II. Molybdenum. Proc Symp paddy soils. Science Press, Beijing
Thompson JE, Legge RL, Barber RF (1987) The role of free radicals in senescence and wounding. New Phytol 105: 317–344
Tisdale SL, Nelson WL, Beaton JD (1985) Soil fertility and fertilizers. MacMillan, New York
Tiwari KN, Pathak AN (1982) Studies on Fe–Zn interrelationships in rice under flooded and unflooded conditions. J Plant Nutr 5: 741–742
Treeby M, Marschner H, Römheld V (1989) Mobilization of iron and other micronutrient cations from a calcareous soil by plant-borne, microbial and synthetic metal chelators. Plant Soil 114: 217–226
Trier K, Bergmann W (1974) Ergebnisse zur wechselseitigen Beeinflussung der Zink-und Phosphorsäureernährung von Mais (*Zea mays* L.). Arch Acker Pflanzenbau Bodenkd 18: 65–75
Trierweiler JF, Lindsay WL (1969) EDTA ammonium carbonate soil test for zinc. Soil Sci Soc Am Proc 33: 49–53
Trolldenier G (1977) Mineral nutrition and reduction processes in the rhizosphere of rice. Plant Soil 47: 193–202
Ulrich B (1984) Waldsterben durch saure Niederschläge. Umschau 11: 348–353
van Breemen N (1978) Landscape, hydrology and chemical aspects of some problem soils in the Philippines and in Sri Lanka. A terminal report submitted to IRRI. International Rice Research Institute, Manila, Philippines, pp 247–282
van Breemen N, Moormann FR (1978) Iron toxic soils. In: Soils and rice. International Rice Research Institute, Manila, Philippines, pp 781–800
van Mensvoort ME, Lantin RS, Brinkman R, van Breemen N (1985) Toxicities of wetland soils. In: Wetland soils: characterization, classification and utilization. International Rice Research Institute, Manila, Philippines, pp 123–138
Vianello A, Macri F, Bindoli A (1987) Lipid peroxidation induced NAD(P)H and NAD^+ dependent in soybean mitochondria. Plant Cell Physiol 28: 1263–1269
Vlamis J, Williams DE (1967) Manganese and silica interactions in the graminae. Plant Soil 28: 131–140
Wallace A, Cha JW (1986) Influence of iron efficiency in soybeans on concentration of many trace elements in plant parts and implications on iron-efficiency mechanisms. J Plant Nutr 9: 787–803
Xie RJ, MacKenzie AF (1989) Effect of sorbed orthophosphate on zinc status in three soils of Eastern Canada. Can J Soil Sci 40: 49–58
Xie RJ, MacKenzie AF (1990) Sorbed ortho- and pyrophosphate effects on zinc reactions compared in three autoclaved soils. Soil Sci Soc Am J 54: 744–749
Yoshida S (1981) Fundamentals of rice crop science. International Rice Research Institute, Los Baños, Philippines
Yoshida S, Tanaka A (1969) Zinc deficiency of the rice plant in calcareous soils. Soil Sci Plant Nutr 15: 75–80
Yoshida S, Ahn JS, Forno DH (1973) Occurrence, diagnosis and correction of zinc deficiency of lowland rice. Soil Sci Plant Nutr 19: 83–93
Zhang F, Römheld V, Marschner H (1989) Effect of zinc deficiency in wheat on the release of zinc and iron mobilizing root exudates. Z Pflanzenernähr Bodenkd 152: 205–210

Chapter 9

Summary: Breeding Plants for Problem Soils – Current Knowledge and Prospects

A.R. Yeo and T.J. Flowers

9.1 Why Grow Crops on Problem Soils?

The chapters in this book have been concerned with a range of different soil problems that limit plant productivity and with the strategies that might be applied to produce sustainable increases in productivity on these soils. Problem soils can generally be divided into two classes according to whether toxicity or deficiency predominate (although toxic soils are often deficient as well). Of naturally occurring problems, the toxicity of aluminium, iron and manganese are the most widespread (Chap. 7). These are common elements whose availability to plants is associated with acid soils. Factors which lead to soil acidification, such as acid rain and excessive fertilisation, increase the incidence of toxicities by these metals. Many soils are naturally saline, but salinisation caused by irrigation is, and has been historically, one of the most worrying cases of the lack of sustainability of agriculture. On some estimates, up to half of irrigation systems are affected by salination. In contrast, heavy metal toxicity usually affects only small and localised areas such as the vicinity of mines. If saline soils have received the most attention in the book it is because of the interests of many of the authors. Soil toxicities are generally difficult to alleviate by management practices.

Deficiencies of many micronutrients (boron, copper, molybdenum and zinc) are very common – zinc deficiency is, for example, the most widespread micronutrient disorder of crops (Chap. 8). In contrast to toxicities, fertiliser application is commonly used to overcome deficiencies, although often at great cost.

The rationale of the book has been that the utilisation of problem soils, whether natural or anthropogenic in origin, will be necessary in the future. The background, though, is that on a global scale food production has, to date, matched population growth and there would at first sight appear to be an enormous capacity for increase in food production using existing cultivated land (Chap. 1). Based upon the difference between potential and realised yields, this should be sufficient to feed at least the projected human population of 14 billions (Chap. 1). A number of factors detract from this optimistic view. Averages disguise the unequal distribution of population density and food production and the variability of climate. Analysis of recent trends has shown that the parity

Monographs on Theoretical and Applied Genetics, Vol. 21
Ed. by A. R. Yeo and T. J. Flowers

between the growth of food production and of population has faltered; less than half of all developing countries have increased cereal production per capita since the 1980s (Chap. 1).

Money and technology would certainly go some way towards raising actual yields towards potential yields, but how far is uncertain. The resources needed to achieve potential yields on a large scale may simply not be available or may undermine the long-term fertility of the land or have other effects which are increasingly recognised as unacceptable short-term gains. Physical degradation of the soil, land erosion, the breakdown of irrigation through salinity and silting, acidification of soil through the over-use of fertilisers, and the environmental costs of excessive release of fertilisers and pesticides are among the constraints to the long-term realisation of the yields achievable under experimental or maximal husbandry conditions. There are, then, limitations to the intensification of agricultural production, if it is to be sustainable, liable to prevent the exploitation of the yield potential of the crops we have already. If this situation remains true in the long term, then a projected population of 14 billion people could only be supported by some increase in the area under cultivation. However, the problems of taking new land continually and indiscriminately into intensive cultivation are now recognised as well (Chap. 1). Thus there are compelling reasons for maintaining production on land that has been degraded by agriculture rather than abandoning it and moving on; examples would be land that has been made saline or acidic.

Another consideration is that large numbers of people live and farm on poor land, whether the origins of the impoverishment are natural or anthropogenic. One proposition would be to ignore poor land and grow crops on productive land elsewhere, but this carries with it the social problem of bringing urbanisation and unemployment to communities of subsistence farmers. There are compelling social reasons for continuing to utilise poor land.

9.2 Approaches to the Utilisation of Problem Soils

If "problem" land is to be utilised for agriculture, then broadly there are three approaches: change the land to suit our existing crops, modify our existing crops to suit the land, or develop new crops appropriate to the problem conditions.

Soil amelioration is a possible solution to some problems, particularly micronutrient deficiencies which may be overcome by fertilisation (Chap. 8), but at the very least this entails an indefinitely on-going cost. Amelioration is rarely a successful solution to soil toxicities because removal is very much more difficult than supplementation, although there are exceptions – the application of gypsum has proven successful in the reclamation of sodic soils. In some cases an engineering solution is necessary, where waterlogging will occur even with good irrigation management there is no alternative to drainage if the land is to be utilised.

Although various authors have touched upon the possibilities of soil amelioration, the approaches discussed in this book have been concerned predominantly with changing crops and developing new crops to suit the land; improvements that can be made to the plant material to enable the better exploitation of scarce mineral resources and the resistance of soil problems such as the toxicities of iron, aluminium, manganese and salinity.

All approaches involve plant breeding. Opinions differ, however, over the extent to which these aims can be achieved by the plant breeder through the use of visual observation alone, and the extent to which physiological, cytogenetic and molecular genetic approaches might be involved. Opinions also differ over the extent to which an understanding of underlying mechanisms is necessary.

Changing the attributes of crops covers a gamut of approaches from selection within and between existing crop germplasm, through recombination and subsequent selection, to the importation of characters from wild species and, finally, to the domestication of wild species as future crops.

9.3 Selection of Parents and Within Breeding Populations

Screening is an essential tool in the identification of parents and in the assessment of breeding populations for desirable genotypes. This assessment has relied on plant breeders, whose judgement has been the cornerstone of conventional plant breeding. The over-riding problem in assessing plants is a logistic one, and economy and precision are often mutually exclusive – for instance, quantitative evaluation at maturity has to be limited to varieties and advanced breeding lines (Chap. 2). A large proportion of the breeding population has often to be rejected at an early stage when the responses to stressful environments are difficult to isolate. Where selection has to be made upon the performance of the individual, rather than the statistical property of a population, the extent of genotype times environment interaction also decreases precision (Chaps. 3, 7). The extent and importance of environmental interaction may be much greater for response to stress than for selection objectives such as agronomic characteristics.

Where and how to carry out selection are not simple matters. Field sites pose large difficulties so far as some problem soils are concerned. Large population sizes are needed to take account of positional variation within the field in the case of salinity (Chaps. 2, 3), although metal toxicity on acid soils is much more uniform (Chap. 7). Variable climatic factors affect the degree, or even the existence, of the stress. Problem soils can often be used for only one generation of selection each year (Chap. 2), perhaps because they are not irrigated or because the stress being evaluated is season-dependent – for instance, drought. A number of approaches seek to overcome the limitations of field screening on problem soils.

One approach is "shuttling" between one generation of selection for stress tolerance on a problem soil with alternating generations of selection for agronomic characteristics on a good soil (Chap. 2). This ensures that the good agronomic characters are not lost in the early generations, but we have as yet little information about how the characters for stress tolerance behave under these criteria of selection.

Another way is the use, with inbreeders, of rapid generation advance. This entails rapid progress, without selection, through the early heterozygous generations (Chaps. 2, 3) and is capable of conserving the genotype of any F_2 individual capable of setting seed. Although this should ensure that fewest potential lines are lost, logistics precludes the frequent use of the approach.

Another approach is to take initial screening out of the field and into the glasshouse – or even into the laboratory. For instance, the rice germplasm collection has been systematically screened for salinity tolerance in a glasshouse (Chap. 2). The aim is to reduce sample size and year-to-year replication by increasing the inherent reliability of the data. Estimates of heritability of aluminium tolerance based upon screening in culture solution were much greater than for screening in the field (Chap. 7). In more controlled and reproducible conditions, the proportion of the variance in plant response that is due to genetic as opposed to environmental factors is increased. It is possible to make some of these screening techniques simple and rapid enough to use routinely (Chaps. 2, 3, 7). However, individual tests do not generally allow *simultaneous* screening against several stresses where these occur together, and the usual method here is to screen material in target environments (Chap. 2).

All short-term screening procedures use vegetative material. At some stage of a breeding programme, however, screening in the field with plants grown to maturity is essential. Sterility on problem soils in plants that grow well vegetatively is commonly observed, and in wheat the transition from vegetative to generative growth is particularly sensitive to salinity (Chap. 4). On the other hand, it is just not practicable to evaluate every entry in a germplasm collection and to take every cross to maturity in every problem soil. As a consequence, a high proportion of germplasm has to be rejected at an early stage. The approach used by plant breeders has generally been to select for agronomic traits and trust that tolerance to stress remains. This is perfectly logical, but the rather slow progress with breeding for tolerance to soil problems, in contrast to other objectives, suggests that these priorities may not always be appropriate. For example, if rare combinations of characteristics derived from several parents are required or if tolerance can only be achieved by some compromise over the agronomic characteristics of the plant, then the conventional strategy is unlikely to succeed. We suggest that for some breeding objectives the criteria for selection and rejection *have* to be more specific.

Environmental interaction combined with the heterogeneity and unpredictability of environmental stresses are some of the reasons why the overall performance of a genotype or individual may not provide sufficient information on which to judge its worth or usefulness in a plant breeding programme. Other reasons are now becoming apparent from studies giving insight into the mechanisms of tolerance. In the majority of cases, the tolerance of problem soils and associated nutritional disorders has been shown (through either physiological or genetic studies) to be complex. Either a number of different characters are needed together to achieve tolerance, or the same ends may be achieved via a variety of different means (Chaps. 3, 7). Where several traits are necessary, there is unlikely to be progress via any approach which, explicitly or implicitly, treats tolerance as a single character. Any character may be quantitative and any mechanism may or may not be compatible with usual agronomic criteria of success. There is a need to recognise donors that are not only tolerant of a stress, but whose mechanism of such tolerance is not incompatible with agricultural objectives.

Increased understanding of the mechanistic basis of tolerance has fostered the proposal of more sophisticated techniques not only for selecting parental material but generally for screening plants (indeed this is a reason often given in the introductions to scientific papers). The understandable reluctance of plant breeders to share the enthusiasm of laboratory physiologists for new techniques is that, by and large, they cannot be used with breeding populations. For this, a technique must work with hundreds or thousands of individual plants per day, non-destructively, and preferably "in the field"; this frequently means in remote places and with very limited resources. If a technique will not work under these conditions, then it has little credibility as an alternative screening procedure.

If tolerance is conferred by simple characteristics, simple both in ease of recognition and in genetic behaviour, then there seems no reason to doubt that conventional plant breeding can make rapid progress. Where the appearance of the plant is not an adequate guide to performance, however, then the need to improve plants has moved into an area in which our established criteria of plant breeding are no longer sufficient. Techniques giving more information will be needed to characterise plant; they may rely upon physiology and molecular genetics. However, it has to be remembered that such techniques are a means and not an end; they may provide new tools and new criteria, but they are not in themselves plant breeding. There has been repeated reference to a need for an integration of plant breeding, genetics and physiology as the complexity of many of the problems being faced becomes apparent (Chaps. 2, 3, 4, 5).

Lines of approach are (1) to use sophisticated techniques in the selection of parents for crosses but to treat the progeny conventionally, (2) to endeavour to simplify some laboratory techniques to a point where they do have the simplicity and capacity to use in the field and (3) to seek genetic markers for the traits in question (Chap. 3). The rationale of the third is that a single technique, RFLP analysis, could be used for a very wide range of difficult or cryptic characteristics.

9.4 Tolerance and Potential Yield

The question of a penalty in terms of yield potential for the acquisition of tolerance has often been raised. A mechanism of tolerance that allows survival even with very low growth rates can be successful ecologically but is undesirable agriculturally. A reduction in growth rate under optimal conditions is, for instance, associated with tolerance to copper and lead in *Agrostis* (Chap. 6). Not all tolerance to problem soils necessarily involves a yield penalty, however; aluminium tolerance is an example (Chap. 7). A penalty in terms of yield is likely where the plant must allocate considerable resources to adaptation, such as in osmotic adjustment to salinity or to drought (Chaps. 3, 5).

From both agricultural and mechanistic viewpoints it is important to know whether the mechanism of tolerance causes a reduction in yield in the *absence* of stress. In this context it is significant that the effects of some genes involved in tolerance may be constitutive (Chap. 4). It is argued (Chap. 5) that some linkages between tolerance and reduced growth may be very difficult to break; cell expansion after acclimation to osmotic stress has been seen as a characteristic of adapted cells in in vitro culture and this will inevitably have an effect upon yield if it is a characteristic of cells in leaves. It is possible that there is a genetic linkage between adaptation to growth in saline conditions and a slow growth rate, or it may be that a metabolic process that is essential for survival in saline conditions becomes limiting at normal growth rates (Chap. 5). The demands upon the carbon and nitrogen budget made by the synthesis of organic solutes is an example (Chap. 5). The efficient intracellular compartmentation seen in some halophytes of the Chenopodiaceae would minimise this.

A comparison of different halophytes shows that a considerable range of growth rates can be achieved at high salinity according to the mechanism of tolerance (Chap. 3); in some species, growth rates are sufficient for some halophytes to provide economic crops (Chap. 6). Where there is a choice between species, it is important to aim to select a suitable mechanism of tolerance (involving an adequate growth rate) rather than tolerance per se.

A linkage between tolerance and yield potential is not necessarily a problem if the plant material being developed is destined primarily for a problem soil. Thus, a differentiation needs to be made between cases where the soil problem, or other stress, is "inevitable", "likely", or "just possible". If the stress is a rare event, then growing a variety of lower tolerance, but with a higher yield potential than the tolerant variety, is likely to result in better yields most years. If the incidence of the stress is very likely, or inevitable, then a tolerant variety (by definition one that will consistently out-yield the more sensitive variety with the higher potential yield at the levels of stress anticipated) is the clear choice. There will always be a middle position where the frequency or severity of stress falls between these categories and here the choice is likely to depend upon the reason for which the crop is grown; whether for profit or for subsistence. In the latter case, yield stability in poor years will probably be more important than yield potential in good years.

Freedom of choice in terms of the relative tolerance of the variety to be planted is only possible where there is a range of genotypes available to the farmer. The practice on which the green revolution was founded, of release of uniform varieties for planting over large areas, is not appropriate for problem soils. The provision of suitable genotypes requires a shift towards the production of a large number of locally adapted varieties. Current trends in plant breeding for developing countries are towards localised, rather than centralised, varietal development. For instance, the International Rice Research Institute now concentrates on "pre-breeding", the production of elite lines containing characters such as sress tolerance. These are not "finished" varieties; rather they are donors of the characters to be combined with elite local material suited to the other characteristics of the target environment. The final stages of the breeding process, the release of named varieties, is left to local agencies and not attempted by IRRI.

If it is accepted the varieties intended for problem soils do not need to have the yield potential of varieties which would also be grown under more favourable conditions, then some of the emphasis normally placed by plant breeders on agronomic characteristics might be relaxed. For instance, genotypes with greater vegetative stature and lower harvest index might have greater resources available for coping with stress than those normally deemed acceptable in agriculture.

9.5 Genetics of Tolerance to Problem Soils

There are many cases where nutrient accumulation has been associated with a few genes acting in an additive manner (Chap. 2). There are a few cases where a single, dominant gene has been implicated, such as aluminium tolerance in barley (Chap. 7). In wheat, an enhanced discrimination between potassium and sodium is seen as one characteristic associated with salt tolerance, and it is possible that a single structural gene could be responsible (Chap. 4); but many of the problems described have appeared to be complex in their patterns of inheritance. In wheat, for example, the tolerance of aluminium implicates genes on several different chromosomes from the different genomes (Chap. 7) and in other species, control ranges from a multiple allelic series of a single gene to a quantitative trait (Chap. 2). The tolerance of salinity and drought are generally viewed as complex. One reason for this complex behaviour is that tolerance may be achieved in many different ways. In the case of metal toxicity, there are both internal and external modes of tolerance, each of which can be mediated by a number of different mechanisms (Chap. 7). Consequently, the genetic basis of the phenotypic expression of tolerance may not be the same in different lines of the same species, never mind across a range of species.

However, even if several or many genes are involved, it is still possible for the pattern of inheritance to be simple. The possibility of linkage groups of favourable genes means that characters which are separate physiologically may behave in a coherent fashion genetically (Chap. 4). There may need to have been selection pressure during the evolutionary history of a species or its progenitors to give rise to such a grouping, and this may not be the case where there has been insufficient selection pressure to favour the accumulation of several traits into one genotype (Chap. 3). Where a range of different tolerances are concerned, some genes are located on all of the seven chromosome groups of the Triticeae, but group 4 and 5 chromosomes are the most intensely involved and there is a possibility that genes with similar functions are grouped in clusters (Chap. 4).

Tolerance genes may have a constitutive effect, for example leading to low sodium uptake at low as well as at elevated salinities (Chap. 4). There have been few molecular genetic studies of the behaviour of crop plants towards salt stress, but the results there are suggest that stress does not activate genes for tolerance, but that their products are constitutive and it is the "innate genetic make-up that determines fitness to salt stress" (Chap. 4). Work with in vitro culture, on the other hand, does suggest that the expression of some genes is changed (Chap. 5).

9.6 Transfer from Other Species

The crops chosen for domestication and the conditions under which they have been selected reflect the benign conditions during the domestication process, and these crops may not (or at least no longer) possess the genetic variance needed to develop tolerance to some problem soils even if, like saline soils, they are reasonably common (Chap. 6). Species with a wide ecological range can be expected to have a high chance of coming into contact with widespread soil problems, such as acid soils and hence to the presence of toxic concentrations of Al, Mn and Fe. In contrast, the incidence of heavy metal contamination is a rare event ecologically (Chap. 7) and so encountered by few species.

If current genotypes do not exhibit the required degree of tolerance, then some improvement may be achieved by rearranging the genetic variation in more favourable ways, but the upper limit of tolerance is determined by the level of remaining variation. If this is insufficient, then variation may be increased by mutation. This has been very successful in some areas of plant breeding, but has made little impact in developing plants capable of overcoming problem soils. Somaclonal variation in tissue culture is another source of genetic variation. Additionally, selection pressure can be imposed during in vitro culture of cells. The use of regenerants from such selection procedures is only applicable to mechanisms of tolerance that work at the cellular, as opposed to tissue and organismal level, but the fact that tolerant variants can be selected in vitro does suggest that at least some constituents of tolerance do derive from cellular-based mechanisms. One practical difficulty which has been seen with tissue culture is that the frequency of genetic variation, which is one of its main benefits, may undermine attempts to fix desired traits and so reduces its efficiency (Chap. 5).

There are both stable and non-transmissible genetic adaptations: a capacity to tolerate high salinity versus an inherent capacity for adjustment (Chap. 5). Tolerance is sometimes associated with polyploidy (Chap. 5) which provides a stable increased number of copies of a gene. In other cases, it has been suggested that gene amplification in vitro, requiring the sustained presence of salinity, is a reason for the instability of tolerance in regenerated plants.

The other available route to the improvement of existing crops is to seek to increase the genetic base through the importation of genes from related species which grow naturally in the environment to which tolerance is sought. The greatest drawback to hybridisation with wild species as a method of introgressing desired characteristics (assuming it is known which genes are desired) is the large amount of other alien genetic information that is introduced to the genome of the cultivated species. Wild species may have very poor agronomic characteristics and the work needed to regain a useful variety following such a cross will be immense. Introduction of disease resistance genes through wide hybridisation has been successful, but in most cases the poor agronomic qualities of the hybrid prevailed (Chap. 4). As a further example, even though the tomato has been crossed with a salt-tolerant wild species, this has not led to the development of a commercial cultivar. It may be that the tolerance of the wild relative is due to, or linked to, a characteristic (such as a very slow growth rate) that renders the wild genotype agronomically worthless. Thus, it may be that an extreme in tolerance in an ecological sense is not the best choice of parent, but that it is better to compromise and choose a species with less tolerance but better agricultural characteristics (Chap. 3).

The needs are to transfer less of the wild genome, and cytogenetics is concerned with manipulating single, or small parts of a single, chromosome. Though chromosomal manipulation introduces much less unwanted genetic material than hybridisation does, even chromosome substitution and translocation carry alien material as well as the desired genes (Chap. 4). Current aims are to transfer smaller and smaller fragments of chromosome to minimise the introduction of unwanted alien material. The best example of the potential for chromosomal manipulation is demonstrated in wheat (Chap. 4), but a special combination of different factors allows the technique to be successful in this particular case and the general applicability of the approach is less certain. Where sexual imcompatibility makes a wide cross difficult, protoplast fusion has been advocated as a method of bringing about hybridisation. The value of this as a route for the introgression of valuable genes independently of unwanted genetic material has been questioned because of the difficulty of manipulating chromosomal combinations and the inability to select in vitro for agronomically useful genotypes (Chap. 5).

There are two other considerations worthy of mention regarding the import of stress tolerance from wild relatives. The first is the availability of a suitable donor. A good example is the contrast between wheat and rice. In wheat, there are genes which influence tolerance to salt that have been located on the D genome of bread wheat; these genes originated from *Aegilops squarosa*. There are also a number of more distant relatives which may offer better genes for salt tolerance (Chap. 4). In contrast, rice has only one distant, and far from promising, salt-tolerant relative (*Porteresia coarctata*). The second consideration is whether or not the genetic information that is transferred will be expressed in the new genetic background; and this is not always the case. The ranking for aluminium tolerance among the temperate cereals is rye > oats > hexaploid wheat > barley. Attempts to transfer tolerance from rye to wheat by chromosome manipulation have been unsuccessful, and the expected tolerance was not seen in the wheat genetic background (Chap. 7).

Given the problems of wide hybridisations and even chromosome manipulation, the fact that the techniques for transferring one or a few genes between species are becoming increasingly routine is a promising avenue. The fundamental questions associated with moving genes are, however, (1) which genes? (2) from which donor parent? and (3) will they be expressed in the new genetic background? The present situation with regard to problem soils would appear to be that there is very little information about which genes are important and what their products are. One aim of the reductionist approach of examining cells in vitro is to identify gene expression in salt-tolerant genotypes or in genotypes after salt adaptation; the lack of present knowledge about the *function* of any of the gene products is emphasised in Chapter 5.

9.7 Domestication of New Crops from the Native Flora

If genetic variation for tolerance is not readily available in the crop genome and it would be necessary to seek hybridisation or gene transfer from wild species to enhance tolerance in the crop, then an alternative approach is to seek to utilise those wild species as crops in their own right (Chap. 6). For some soil problems (heavy metals), there is not a large native flora to exploit, but for others (high salinity) there is a large resource base of over 1500 species (Chap. 6); this has received much attention. Given the choice between improving the salt tolerance of an existing crop and improving the agricultural characteristics of halophytes, then it is argued (Chap. 6) that it is the latter approach at which plant breeding has been singularly successful.

Yields of many halophytes as forage crops are good by agricultural standards, though crude protein may be only half that of alfalfa (Chap. 6) and chemical constituents designed to deter herbivores do constitute a problem with the use of wild species. At present, a blend to supplement other feed material appears to be the best practice (Chap. 6). Growing halophytic crops would allow the use of saline water to irrigate land, and saline water is often in abundant supply where fresh water is not (Chap. 6). The use of the seeds of halophytes directly as food crops also holds possibilities and here, because of the selectivity of phloem during seed filling, the problems of high salt content found in the harvested vegetative material is absent. *Salicornia* has yielded 2 tonnes per hectare of seed which has a high protein content and very valuable characteristics as an unsaturated vegetable oil (Chap. 6).

The only estimate of possible time scale is derived from the hybridisation of wheat and rye to form *Triticale*. Although both parents were domesticated to start with, the first hybrids had the numerous undesirable agronomic characteristics (tall, late-maturing, photoperiod-sensitive and partially sterile) which might be anticipated in a wild hybrid. It has taken 25 years to make this crop agriculturally useful and progress with "wild" parents would not be quicker (Chap. 6).

9.8 Outlook

This book has described what success there has been with developing plants for problem soils and has highlighted the extreme difficulties being encountered in this area of plant breeding. Many of the reasons for the limitations to progress have been described. These include the lack of suitable variation within some crop genomes and the factors that undermine our established reliance upon the appearance and overall performance of plants as the sole criteria of selection. Because these limitations have been recognised, there has been a repeated call for more integration between plant breeding, genetics and physiology to provide new approaches. A number of ways have been proposed in which these other disciplines could combine with plant breeding; through the introduction of genetic variation by novel means, and through the development of screening and selection criteria tailored to the special needs of developing varieties for problem soils.

There is an understandable expectation that the application of new methods and technologies might lead to rapid advances; but it has to be remembered that, however novel or sophisticated, these methods and techniques are not more than new tools so far as the end result is concerned, which is the development of new cultivars. With conventional breeding and an inbreeding species then, from start to finish, 10 years is rapid for development of a cultivar. The experience with *Triticale* gives some guideline as to how long any unconventional procedure might take. Those who work in disciplines such as physiology and molecular genetics, and those who fund them, will have to view their work in the time scale of plant breeding if progress is to be made.

Subject Index

ABA
  and osmotic adjustment 108
acclimation
  of cells to salt 86
  non-heritability of 87
acid soils 145
  soil-based screening 150
addition lines
  crossing procedure for 68
  in wheat 63–65
agricultural intensification 202
agricultural sustainability 202
Agrostis
  ecological amplitude 154
Agrostis capillaris 153
alien DNA 76
alien gene transfer 66, 76, 78
  and *Thinopyrum bessarabicum* 67
allele frequency 166
aluminium plus manganese 146, 147, 151, 153, 154, 155
  genetics of tolerance 160–164
  and species distribution 154
aluminium tolerance
  calcicole/calcifuge 153
  chromosomal location 161
  in crops 155
  expression in wheat 161
  genetic control of 18, 163
  genetic variation for 153
  genetics in wheat 161
  heritability 162
  intraspecific variation 156
  in maize 157, 158, 162
  in *Phalaris* 158, 163
  physiological basis 164
  problems in measurement 163
  in rice 157, 163
  in rye 161
  screening 149
  in sorghum 157, 161, 162
  transfer from rye to wheat 161
  in wheat 155, 156
  and yield penalty 158
amphiploid (amphidiploid)
  salt tolerance of 69
  wheat/*Thinopyrum bessarabicum* 67
  wheat/*Thinopyrum elongatum* 70
ATPase 96
Atriplex
  cell culture 104
  as forage crop 132
  glycinebetaine in 104, 105
  improvement of 137, 138
  nutritional value 134

backcross breeding 26
barley
  salt tolerance of 72, 73–75
boron
  availability . 191
  occurrence 191
  roles in the plant 191
boron deficiency 192
  occurrence 191
boron toxicity 159
  occurrence 192
  symptoms 192
breeding methods
  heterosis 28
  mutation 27
  pedigree 23–25
  physiological 38–43
  polyploid 27
  pure line 21, 22

calcicole 153
calcifuge 153
calcium
  ion channels 98
  transport 97
cell culture 83–111
  tobacco 106, 107
Chenopodiaceae 101, 104
  as saline crops 134
Chinese Spring 71
  and aluminium tolerance 161

chloride
  transport 98
chromosome manipulation 61–78, 209, 210
chromosome transfer
  small fragments 76
combining ability
  and aluminium tolerance 162
compartmentation 47, 94, 95
  and metabolism 47, 48
  regulation of 96
compatible solutes 99, 109
copper
  availability 189
  occurrence 189
  roles in the plant 189, 190
  solubility 189
  uptake by plants 189
copper deficiency
  interaction with zinc 190
  occurrence 190
  symptoms 190
crops
  conditions during domestication 127
  ecological range of 39
  evolution of 127
  native flora as 131–140
  species used as 9, 39
cybrid 90, 91
cytogenetics 61–78, 209

D-genome (of wheat)
  and salt tolerance 62
Distichlis
  as fodder 132
Distichlis palmeri
  as halophytic seed crop 136
DNA
  insertion 91
  transfer 91
domestication 208, 210, 211
  in crop development 131–140
  stages in 137
  timescale 137, 139
drought 49–54
  breeding for tolerance 43
  complexity of tolerance 49
drought tolerance
  criteria for 53, 54
  and root morphology 51, 53
  and root/shoot signals 52

ecotypes 146

flowering time
  and salt tolerance 71, 72

food supply 2–7
  actual and potential yields 202
  and area harvested 4, 5
  and diet 4
  and human population 2–5
  and unit production 5
  per capita 2, 3
  potential for increase 7
  projections 3–6

gene(s)
  clusters 75, 208
  extranuclear 91
  linkage 20, 208
  linkage and breeding population 20
  location on chromosome 19, 67, 71, 72, 74
  and metal tolerance 130
  in salt tolerance 67
  transfer 91, 92, 129, 210
genetic background 69
genetic control 16, 17, 18
  of metal toxicity 17, 18
  of mineral efficiency 16, 17
  of mineral uptake 16, 17
  of salinity tolerance 17, 18
genetic markers 76, 77
  and indirect selection 76, 77
Glennson 81 70, 71
glycinebetaine
  adaptation to osmotic stress 108
  as compatible solute 101
  deficient mutants 102
  maize hybrids 102, 103
  pool size 104
  regulation of 101
  salinity and synthesis of 103
  synthesis of 105

halophytes 45, 104, 127–140
  agronomic improvement 139–140
  as crops 62, 210
  farming of 138
  as food crops 135–136
  as forage 132, 135
  nutritional value 134, 135, 136
  palatability 134
  productivity as crops 133
  resource base 131
  and saline irrigation 133, 210
  as seed crops 136–137
heavy metal tolerance 129
  agricultural relevance 129
  enhancers 160
  genetic resource base 130
  major genes 159

physiological basis 164
polygenes 160
in wild species 152
heavy metals
binding 130
and natural selection 146
hematoxylin
in aluminium screening 149
heritability
of aluminium tolerance 163
heterogeneity
aluminium and manganese in field 150
heterosis breeding 28
Hordeum
salt tolerance in 72
Hordeum spontaneum 75
human population 1, 2
carrying capacity 5
current and projected 1, 2
distribution of 1, 2
growth rate of 1
hybrid crops 71

introgression 83
ion channels 96, 97, 98
activation by calcium 98
carriers 96
gating of 97
iron
availability 176, 179
effect of flooding 176
interaction with phosphate 179
interference with uptake 177
mechanisms of uptake 177
minerals of 176
occurrence 176
roles in the plant 178
solubility 176, 179
iron deficiency 175
amendment 179
occurrence in rice 178
symptoms of 178
iron toxicity
amelioration 182
causes 181
occurrence 180
and oxygen radicals 181
symptoms 180
isozyme markers 77

Kosteletzkeya virginica
as halophytic seed crop 136

land use 4–7
and agriculture 6
animal crops 4, 7
potential arable 6
and soil problems 6, 7
sustainability 5, 7, 8
Leptochloa
as fodder 132
local adaptation 43
Lycopersicon
and wide crossing 40

maize
aluminium tolerance 157, 158
and glycinebetaine 102, 103
manganese
availability 182
occurrence 182
roles in the plant 182
uptake of 182, 183
manganese deficiency
amelioration 184
effects of 183
occurrence 183
symptoms 184
manganese tolerance 159
in Australian pasture legumes 158
mechanisms 185
manganese toxicity 150
occurrence 184
symptoms 184
mass selection 22, 23
metal (*see also* heavy metal, aluminium, manganese)
metal tolerance 39
criteria 146
and crop improvement 146
evaluation of 147
evolution of 153
field and laboratory methods 151
genetic basis 159
measurement of 147
soil-based screening 150
unconscious selection 156
metal toxicity 145–168
anthropogenic 145
impact on agriculture 146
natural 145
sources of 145
metallophytes 130
metallothioneins 130
micronutrients 175–193
mine waste
colonising species 152
mineral efficiency 16, 129
genetic control of 16
and wild relatives 129

molybdenum
  availability 193
  occurrence 192
  roles in plant 193
molybdenum deficiency 193
  occurrence 193
mutation breeding 27

native flora
  agronomic use of 127–140, 210

organic solutes 99, 100
osmoprotectant 100
osmotic adjustment 52, 53, 94, 99–110
  and ion accumulation 94
  and salinity 47
osmotic solutes 99–110
osmotic stress tolerance 94–110
  molecular determinants 92

palatability
  and herbivore resistance 135
  and salt content 135
pedigree breeding 23–25
  bulk method 24
  modified bulk method 24, 25
  and rice salt tolerance 24
  shuttling 25
Phl gene 66, 67, 76
physiological criteria/characters 37–57
  in screening 37, 41
physiological screening 54–57
  in breeding populations 55, 56
  and genetic markers 56
  in line evaluation 55, 56
  in parent selection 55, 56
phytosiderophores
  and iron uptake 177
plant introductions 21
polyploid breeding 27
potassium
  transport 97
problem soils
  acidity 201
  amelioration 202
  complexity of 37
  and crop improvement 203
  deficiencies 201
  discontinuity 38
  and food supply 201
  reasons for use 202
  toxicities 201
  utilisation 201
proline
  accumulation and osmotic stress 105
  in cell cultures 106
  and compartmentation 104
  cost to N-budget 107
  expansion flux 107
  pathway of biosynthesis 109
  regulation of pool sizes 106
  regulation of synthesis 108
  in root apex 105
proton pumps 96, 100
  regulation 99
protoplast fusion 89, 90
  limitations 90
pure line breeding 21, 22
  and rice salt tolerance 22
pyramiding 37, 85, 89
pyrophosphatases 96

rapid generation advance 27, 204
recurrent selection 26
RFLP 56, 77, 205
rice 11–15, 21–29, 175–193
  aluminium tolerance 157, 163
  crop statistics 175
  micronutrient deficiencies 175
  root oxidation 181
root growth
  in aluminium screening 14
  and metal tolerance 147
  and metal toxicity 147
  regrowth 149
  in screening methods 148
  tolerance index 148, 149
root oxidation
  and iron 181

Salicornia
  as oilseed crop 137
Salicornia bigelovii
  as halophytic seed crop 136–137
saline irrigation
  drainage 138
  frequency 138
  leaching 138
  management 138
salinity (*see also* salt tolerance) 43–49
  avoidance mechanisms 45
  basis of damage 44
  and drainage 133
  extent of saline land 62
  limitations of salt exclusion 44
  and osmotic adjustment 45
  salt exclusion 44
  target environments 46
salt marshes 109, 110
salt tolerance 61–78, 83–111, 127–140
  adaptation and growth rate 110
  and agricultural production 89
  and cell expansion 88, 89
  cell lines 86

cellular mechanisms 88
complexity of 48, 49, 128
criteria for 86
and flowering time 71, 72
genetic variability for 128
and glycinebetaine 101
heritability in cell lines 87
in vitro and in vivo 85
inheritance in cell lines 88
and innate makeup 71
loss of genes in domestication 139
molecular determinants 92
plant growth 93
pleiotropic effects 71
polyploid cell lines 87
potassium: sodium selectivity 95
selection criteria 88
stability in cell lines 87
in Triticeae 62–78
and vigour 48, 73, 88
of wheat 62–78
wild relatives 128
salt uptake
control of 46
damage limitation 46
dilution by growth 48
mitigation of 45, 46
saponins 135
screening (*see also* selection) 11–14, 41, 149–151
and allele frequency 167
applicability of laboratory methods 205
design 41
environmental interaction 203
field techniques 11, 14, 203
genotype/environment effects 204
glasshouse techniques 11, 14, 204
laboratory techniques 11, 14
logistics 41, 203
and mechanism 167
for metal tolerance 152, 165, 166
methods 39, 49, 54, 55
physiological techniques 41, 54, 55
precision 203
procedures for rice (tabulated) 12, 13
root growth techniques 147, 149
tolerance vs agronomics 204, 207
vegetative/generative stages 204
visual techniques 41
selection (*see also* screening) 149–151
for metal tolerance 165
physiological 54–57
shuttle breeding 204
single seed descent 27
sodium
antiport 97
antiport at tonoplast 98
transport 97
somaclonal variation 84, 85
limits to exploitation 85
stress(es)
complexity 42
interaction 42
spatial/temporal distribution 38
stress tolerance
agricultural/ecological advantage 45, 47, 48
and ecological range 39
complexity of 38, 39, 204, 207
component traits 41
constitutive expression 206
ecological/evolutionary background 37
genetic basis 207
genetic variability in crops 128
interspecific variability 15
intraspecific variability 15
linkage with reduced growth 206
measurement of 41
to metals 39
multiple adaptations 38, 39
variability in rice 16
yield penalty 206
succulence 47
sulphide
detoxification of 109

Thinopyrum bessarabicum 67, 69, 70
tissue culture 83–111
limitations of 110
tobacco
cell culture 106, 107, 108
tolerance (*see* stress tolerance and individual stresses)
transpiration
and canopy temperature 53
Triticale 71, 128, 138, 139, 211
turgor 53

variation
free and potential 166
from cell culture 208
from mutation breeding 208
limitations in domesticated species 208
sources of 208
vegetative propagation 85
vigour 48
in salt tolerance 73

water use
and yield 50

water use efficiency
and gas exchange 50
and respiratory losses 51
and salt tolerance 49
as a selection criterion 51
and stable isotopes 51
wheat
addition lines 63–65
aluminium tolerance 155, 156
chromosome complement 63, 66
genomes of 63, 66
monosomic series in 63
ploidy 63, 66
recombination in 66
substitution lines 63, 64, 65
translocation lines 66
wide hybridisation (crossing) 39, 40, 75, 209, 210
agricultural vs ecological advantage 40
and cell fusion 90
and salt tolerance 67, 128
wheat/alien hybrids 63–65, 68–71
wild species relatives as donors 62, 209, 210
heavy metal tolerance 152
yield 37, 41
and crop water use 50
limitation as selection criterion 42, 43
and osmotic adjustment 53
records and averages 7, 8
stability 42, 43, 88
yield penalty
and aluminium tolerance 158
and target environment 206

zinc
availability 185, 187
contamination of soils by 155
crop tolerance 155
occurrence 185
requirement by rice 188
roles in the plant 185, 186
solubility 185
uptake of 188
zinc deficiency
alleviation 188, 189
effects of flooding 187
effects of phosphorus 187
occurrence 186
symptoms 186

MIX
Papier aus verantwortungsvollen Quellen
Paper from responsible sources
FSC® C105338

If you have any concerns about our products,
you can contact us on
ProductSafety@springernature.com

In case Publisher is established outside the EU,
the EU authorized representative is:
Springer Nature Customer Service Center GmbH
Europaplatz 3, 69115 Heidelberg, Germany

Printed by Libri Plureos GmbH
in Hamburg, Germany